Cranberry Harvest

Cover photograph: The Cranberry Harvest, Nantucket Island, *painted in 1880 by Eastman Johnson. From the Putnam Foundation Collection at the Timken Art Gallery in San Diego, California.*

Below: Scoop-harvesting on a Plymouth County bog, ca. 1910.

Cranberry Harvest

A History of Cranberry Growing in Massachusetts

Joseph D. Thomas, Editor

Spinner Publications, Inc.
New Bedford, Massachusetts

Special Thanks to

The Cape Cod Cranberry Growers' Association
Ocean Spray Cranberries, Inc.
The University of Massachusetts Cranberry Experiment Station
The Middleborough Public Library
The Plymouth Savings Bank

First Edition

Credits

Design and Production   Joseph D. Thomas

Editing   Dianne Wood

Photography and Design   John Robson

Writers

Carolyn C. Gilmore
Christy Lowrance
Constance Crosby
Robert Demanche
Megan Tarini
Linda Donaghy
Marilyn Halter
Dan Georgianna
Linda Rinta

Contributors

Larry Cole
Beverly Conley
Lizanne Croft
Robert A. Henry
Elaine Lembo
Ted Polumbaum
Michael Zaritt

Funding received from

The Massachusetts Council on the Arts and Humanities
The New Bedford Arts Council

We would also like to thank those who took the time to talk with us, share their photographs and make their resources available:

Steve Ashley
Joseph Barboza
Karen Barnes
Mary Barros
Doug Beaton
Elliot Beaton
Joseph Brigham
Walter Cannon
Jeff Carlson
John Clark
Skip Colcord
Orrin Colley
Manuel Costa
Shirley Cross
Nancy Davison
George Decas
John C. Decas
Irving Demoranville
Albertina Fernandes
Jean Gibbs
Kirby Gilmore
Clark Griffith
John Hall
Robert Hammond
Arthur Handy
Eino Harju
Paul Harju
Wilho Harju
Irving Howes
Jim Jenkins
Robert Johnson
Carl Johnson
Marjorie Judd
Charles Kallio
Seth Kallio
Ricky Kiernan

Edwin Korpinen
Mary Korpinen
Wilho Lampi
Francis LeBaron
Lucillia Lima
Maurice Makepeace
Chris Makepeace
Donald Malonson
David Mann
David Mendes
Steve Moniz
Flora Monteiro
George Olsson
Dwight Peavey
Francis Phillips
Gertrude Rinne
Jack Saaramaki
Robert St. Jacques
Philip Sullivan
Linc Thacher
Jane Thomas
Paul Thomas
United Data, Inc.
Gary Weston
Gladys Widdiss
Ted Young

A.D. Makepeace Company
Cape Cod Community College
Decas Cranberry Company
Massachusetts Horticultural Society
New Bedford Public Library
Sandwich Public Library
The *Standard-Times*
Sturgis Library
Wareham Public Library

Foreword

Cranberry Harvest began as a project to commemorate the 100th anniversary of the Cape Cod Cranberry Growers' Association. We hope it will serve to commemorate the achievements of everyone who has "worked the bogs" of southeastern Massachusetts.

The cranberry bogs have always been special to me. As a young man, I worked two picking seasons for an elderly Carver grower named Al Eastman and lived in his bog shanty not far from the Edaville tracks. With a potbellied wood stove for heating and cooking, an outdoor hand pump for water, and a small cot to sleep in, I was in heaven. It was a time, I felt, that the more primitive the conditions, the better for character-building and for my spiritual well-being.

With a crew of four, a Model A bog buggy and three Western pickers, we harvested the small bogs in the crisp autumn air. One person handled the buggy, picked up the burlap bags full of berries and brought them to the screenhouse (a barn). There we screened berries until the sun went down. When the day was done I returned to the shanty where it took the rest of the evening to get the stove going, heat up water, eat and clean up. Maybe there was time to read a little Thoreau by lantern light.

Working the bogs was a brief work experience, but one that gave me an appreciation of the land, the work and the people associated with cranberry growing. And when I'm on the bogs, I am still moved by the sweet scent of pine and cedar, of damp earth seasoned with decaying vines and thriving plants. For me, it is a reminder of the good things in life.

From the time I began working with Spinner, I always wanted to publish an account of life on the cranberry bogs. So when Clark Griffith and other members of the growers' historical committee approached Spinner about doing this history, I was thrilled. The only problem has been knowing when to let it go.

As I reluctantly lay this project to rest, I have many people to thank. Especially my wife Jane who has worked closely with me and has put up with my intensity.

There are many people who have given immeasurably to this publication. My co-producers John Robson and Dianne Wood, and free-lance writers Bob Demanche and Dan Georgianna, have worked far beyond their commitments to make this a quality production. I am also grateful to Dick Carroll, Pat Cunning and the Plymouth Savings Bank for their confidence and consultation, and to Carver grower Clark Griffith for his support.

And a very special thanks to Mr. Larry Cole, cranberry grower, historian, and outspoken ambassador representing cranberry growers everywhere. Larry's devotion to his work is only exceeded by his generosity. From the very start, he has made available all of his research and writing materials (which he had hoped to publish), and he has traveled long distances to be interviewed or to give advice. He has opened doors and opened his heart to us in every way and I am forever grateful.

Joseph D. Thomas

Contents

Introduction

The Cape Cod
Cranberry Growers' Association

Leading the Way

Carolyn C. Gilmore

When the Cape Cod Cranberry Growers' Association was formed in 1888, its membership was concerned about a standard barrel size, uniform pricing and better marketing of the crop. The men who led the way, including the first presidents—John J. Russell, Abel D. Makepeace, Emulous Small and George Briggs—were risk takers and entrepreneurs. Theirs was the first generation of growers to expand cranberry plantings from backyard swampholes into large commercial family farms, because they realized a potential for extensive markets for the native fruit. They also saw the need to unite as a means to this end. As the collective voice for Massachusetts growers, the Cape Cod Cranberry Growers' Association accomplished what an individual couldn't do alone.

Cranberry culture first took hold on Cape Cod following Henry Hall's experiments in 1816 on transplanted wild vines in his North Dennis cranberry yard. For the next 30 years cranberries were grown as a supplement to other forms of agriculture and income.

Captain Zebina Small of Harwich was an early grower. He lost $400 on his first cranberry venture in 1847, probably because he tried to grow his vines submerged in water. He then studied Captain Cyrus Cahoon's growing techniques, and on his second try he was successful. Twenty years later Small was a knowledgeable and well-established grower.

In 1866, Small and two other prominent growers, Obed Brooks and Nathaniel Robbins, advertised for a Cranberry Growers' Convention "to consider the best method of cultivation and other such matters." Almost 70 growers from Barnstable County gathered for a full-day meeting and formed a group called the Cape Cod Cranberry Growers' Association. Although this group is not generally considered the beginning of the present association by the same name, it set the stage for communication as a united forum for growing better crops and expanding the cranberry market.

This association, presided over by Small, drafted a constitution, agreed on a 110-quart standard barrel and hired an entomologist to investigate and make recommendations on growing problems. Later Small took it upon himself to tour New England and New Jersey cranberry growing areas just before the harvest to examine the crop prospects and share his findings with Cape Cod growers.

This first Cape Cod Cranberry Growers' Association became defunct after only a few years. However, a generation after Small's death, 100 cranberry growers met again in Sandwich for a "Cranberry Convention" in July 1888. Out of that meeting a second Cape Cod Cranberry Growers' Association was formed. The local newspaper, the Cape Cod Item, described the meeting as "a spontaneous movement with no personal object beyond a better understanding among the growers, the maintaining of the high character of the fruit and square dealing with the public." The group approved a constitution that defined the association's mission "to promote the interest of its members in whatever pertains to the growth, cultivation and sale of cranberries."

At their charter meeting, the membership agreed to adopt a standard size cranberry barrel to eliminate confusion in the marketplace, just as the earlier association had in 1866. It also looked into setting up a "trademark" for its membership, but A.D. Makepeace found that legally it could not be limited to their group. Trademarks were, however, attached to the marketing organizations

these early growers formed at the end of the century.

The association's charter members were men who envisioned a future for large-scale cranberry plantings. They took mortgages and risked capital in pursuit of this vision. Already available were lands stripped for other purposes. The iron and steel trades in Plymouth and Cape Cod were on the way out, leaving behind shallow open pit mines and mill ponds once used for powering the smelting furnaces. And the natural Atlantic white cedar swamps had been logged over for posts, railroad ties and shingles. Using the wastelands of other industries, the early growers slowly learned to coax a crop from these man-made wetlands. Thus the cranberry industry moved out of the backyard into the twentieth century.

John J. Russell, the association's first president, was chief financial officer of the Plymouth Savings Bank when he became involved in the fledgling cranberry industry. The bank had a policy of not making cranberry loans and denied grower George P. Bowers a loan to finish a large bog he had started on the East Head Stream on the Plymouth/Carver line. Russell, who was a personal friend, lent Bowers the money to finish the project and later took back a four-acre bog in payment.

Abel Dennison Makepeace served as the association's next president, from 1893 until 1895. He began business as a harness maker and saddler in Middleboro before moving to Hyannis at the age of 22 to farm strawberries and potatoes. He was among the pioneers who before 1860 had lost money trying to build commercial bogs. In 1874 he obtained the financial backing of George Baker of Boston and bought a large tract of land in Newtown in Barnstable. Later he acquired water rights in Plymouth County, where he would build his largest bogs. By the start of the association in 1888, his crop was up to 16,000 barrels. The cranberry business was becoming recognized as a solid investment, and he was invited to serve as a director of the Hyannis National Bank.

Through the years, the association has served the needs of both large and small growers. Even today, more than half its members have ten acres or less of cranberry bog. It has also functioned as a training ground. Leaders of the association have often pioneered efforts requiring cooperation and compromise in marketing organizations.

Marcus Urann, president from 1929 to 1931, observed that among growers "Cooperation is a way of life and those who believe in it should work together rather than work in several individual groups, no one of which can be strong enough to do a real job."

10

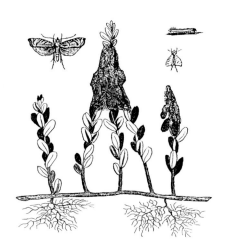

A favorite topic at association meetings is insect and disease control. At one early meeting, growers including George Miller, Franklin Crocker, Ambrose N. Doane and H. H. Heald discussed "The Worm, Its Ravages, Prevention and Cure." It is not surprising, then, that professional entomologists came to address the growers from time to time. J. B. Smith, professor of entomology at New Jersey State Agricultural College, was the first to speak on insect problems in 1899. The outcome was a committee appointed to work on securing a cranberry research station for the Commonwealth. This became a reality in 1910. The station's first director, Henry J. Franklin, had addressed growers for the first time in 1906, thus beginning his 46-year association with Massachusetts cranberries. After his first season on the bogs he discovered the life histories of fireworm, false army worm and yellow and black headed cranberry worms, and suggested controls for these insect pests.

Growers also tried to settle industry-wide labor shortages by agreeing on a payment policy for pickers. In 1892 members debated "a long time" before adopting a policy of 9 cents per six-quart measure. Price regulation was taken up by the growers in 1901, the year of a short crop. Growers decided to fight low prices by not selling the crop for less than $6 a barrel or before September 20.

During the aminotriazole scare and subsequent market crash of 1959, an informational "mass meeting" hosted by the association was attended by 600 growers. There a panel discussion on obtaining government indemnity for the condemned crop and ways to restore consumer confidence was facilitated by president Gilbert Beaton.

Publicity was also a topic of the association over the years. A small recipe book was suggested in 1904 to encourage housewives to make better use of cranberries. In 1936 Russell Makepeace injected some humor into a meeting by reading an 1808 report by a French salesman working in Boston trying to "induce Americans to eat more French products." This man found Americans to be "barbarians in their eating habits" and blamed the situation on "the eating of too much cranberry sauce."

The Cape Cod Cranberry Growers' Association of 1990 faces different challenges for the continuation of the family farm in an increasingly urban environment. "We're working to protect the right of the cranberry grower to grow cranberries and manage his land," said president Jeffrey Kapell. "We ourselves are environmentalists, working with land systems and maintaining the quality of the environment."

The association has become a watchdog for growers in Massachusetts, petitioning policy makers at the state and local levels on their behalf. By doing so the association has given the growers both a single voice and a collective strength.

Opposite page: "Making A Cranberry Bog," Harper's Weekly, October 10, 1885. *Describing the essentials and risks of cranberry culture, the article suggests, "A farmer . . . will mortgage his farm to put in this source of revenue." At a cost of $350 an acre, and with an average yield of 100 barrels per acre, after three years, ". . . the bog is expected to bear and to pay all expenses of construction." (Courtesy of George Decas)*

Left: "Vine worm and its work." The dreaded vine or "fire" worm is shown as larva and moth. It was a wretched enemy of the early grower and one reason why growers felt it necessary to organize. Engraving from Cranberry Culture *by Joseph J. White.*

Below: Exhibiting at trade shows, agricultural fairs and festivals has always been an important function of the growers' association. At this 1930s New England Agricultural Fair, Massachusetts growers showed off cranberry products and the techniques of their industry. (Courtesy of Nancy Davison)

Part One

From Swamps to Yards

Overview

Christy Lowrance

Despite its sturdy, simple appearance, the cranberry is a fussy little plant. Although it grows wild from the Carolinas to the Maritime Provinces of Canada, it has quite specific environmental requirements, and at one time only 25,000 acres in the United States were considered suitable for its cultivation.

Small wonder, then, that Cape Cod is the birthplace of the cranberry industry, for its glacial characteristics have endowed it with every ingredient necessary to the prosperity of a cranberry plant: acidic peat soils, coarse sand, a constant water supply and a moderately long and frost-free growing season.

An indigenous inhabitant of the Cape and southeastern Massachusetts, the cranberry was valued by Native Americans centuries ago for its medicinal and nutritional properties. They used it in pemmican, a meat and fat mixture, as a dye and as a poultice for wounds. The Pilgrims found it not only edible, but "excellent against the Scurvy…and to allay the fervour of Hot diseases."

There were a number of places on the Cape where Indians gathered wild cranberries, including Waquoit in Falmouth, Popponesset and Santuit in Mashpee and Sandy Neck in Barnstable. According to Hazel Oakley, genealogist for the Wampanoag Tribe, Native Americans were industrious farmers, but they did not cultivate cranberries because the natural supply met their needs.

One ancient Indian legend gives the cranberry a miraculous role in saving the life of Richard Bourne of Sandwich, an early Christian missionary and friend of the Indians. It describes an encounter between Bourne and a medicine man, who unkindly cast a spell that caused Bourne to become mired in quicksand. The immobilized missionary challenged his foe to a battle of

wits and promised that if he lost, he would serve the medicine man, but if he won, he must be freed and "troubled no more." Day after day the contest continued, with Bourne apparently none the worse for wear, though the medicine man steadily weakened.

Unnoticed by the medicine man was a white dove that periodically flew down from heaven and placed a round red berry on Bourne's lips. Once the dove dropped its berry, which rolled away and became embedded in the riverbank nearby. When the medicine man spotted it, he realized that the dove was feeding Bourne and tried unsuccessfully to cast a spell on it. Eventually, he conceded the match, saw the evil of his ways and converted to Christianity. Presumably, Bourne was troubled no more.

Meanwhile, the berry that the dove let fall grew and fattened by the river. Finding it there, the Cape men knew there was truth in

Above: A cranberry box label showing a map of Cape Cod and Plymouth County. The natural sandy soil and rich peat layer of this region are perfect for cranberry cultivation, as is its topography of broad swamps linked by wandering brooks and intermittent streams. (Courtesy of Larry Cole)

Opposite page: The American cranberry, Vaccinium macrocarpon, also known as "the little waif of the swampland," the "blushing belle of the winter board" and "ruby of the bog." The horizontal stem is called the runner. The uprights, growing out of the runner, bear the fruit. This is the cherry variety, so named for its cherry-like shape. From Joseph J. White, Cranberry Culture, 1870.

the story that the cranberry came down from heaven in the beak of a winging dove. (*The Narrow Land*, Elizabeth Reynard)

By practically all accounts, cultivation of the cranberry was first attempted by Henry Hall of Dennis around 1816. A number of other individuals on the Cape, including Thomas Hall, Elkanah and William Sears, Captain Zebina Small and Asa Shiverick, as well as farmers in other Massachusetts counties and New Jersey, were exploring the possibility of cranberry growing at about the same time. But Hall's role as the pioneer of cranberry cultivation remains unchallenged.

Henry Hall, a veteran of the Revolutionary War and captain of the schooner *Viana*, was born in 1760 or 1761 in Nobscusset (now North Dennis). He was a descendant of John Hall, who emigrated from England to America in 1630.

In the early years of the 1800s, Henry and two others established a salt works on public land near Kiah's Pond, where wild cranberries had always been gathered. The low, swampy acreage was covered with water in the winter and dry in summer, and Hall observed that when sand from a nearby knoll blew onto the cranberries, they flourished rather than being smothered.

Hall fenced in the area to protect it from his cattle, and when the plants thrived, he decided to transplant them, despite the ridicule of his neighbors. Once it was evident that the plants could survive transplanting, he began to clear and drain other areas of his property for what he called his "cranberry yards."

By 1820 Hall had enlarged his cranberry yards enough to produce 30 barrels of cranberries, which he shipped to New York for sale. In 1832 *The Naturalist* reported his yields to be between 70 and 100 bushels a year per acre:

> Mr. Hall practices taking the plants from
> their natural situation in autumn, with balls
> of earth around their roots, and setting them
> three or more feet distant from each other.
> In the course of a few years they spread out
> and cover the whole surface of the ground,
> requiring no other care thereafter, except
> keeping the grounds so well drained as to
> prevent water from standing over the vines.

For whatever reason, Hall didn't expand on his success. In 1834 he deeded his yards to his son, Hiram, who carried on and became quite successful. Nevertheless, Henry Hall's work was sufficient to win him a unique place in the history of the cranberry industry. Clarence J. Hall, in a 1949 *Cranberries* magazine article, said:

> The evidence seems overwhelming that…
> the flame of cranberry cultivation, for one
> reason or another, was kindled at Dennis by
> Henry Hall and his neighbors.

Henry and Hiram Hall have the dubious distinction of being the first men on Cape Cod to pay taxes for cranberry land. They were assessed $4.80, with Henry paying one-third and Hiram two-thirds. As previously unusable marshlands on Cape Cod were developed for agriculture, the potential for revenue was not lost on town officials, and taxation for cranberry bogs began to appear regularly in town records.

There was no particular momentum to cranberry growing until around 1846, when Capt. Alvin Cahoon, who had seen Hall's vines, planted his own bog in the Pleasant Lake area of Harwich. In a statement given in 1851, Cahoon said:

> In the spring of 1846, I cleared off the brush
> from about seven rods and set it with cran-
> berry vines in hill 18 inches apart each way.
> The first and second years, the vines grew
> well and bore a little. The year past the
> average crop was one and one-quarter bush-
> els per acre; more set in 1847 yielded six
> bushels; in '50, 25 bushels; in '51, 54 bushels.

By 1848 Cahoon had about one and a half acres planted and four more cleared, but further expansion was impeded by the high water level in Seymour's Pond. Two hundred yards away was Hinkley's Pond, or Pleasant Lake, which was lower than Seymour's. Cahoon decided to drain the higher pond into the lower one, and with the help of his two young sons and occasional hired men, he hand-dug a 5-foot ditch through 30-foot hills between the two. Using shovels and wheelbarrows, they

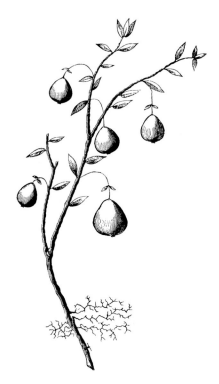

An engraving of the Bell cranberry variety, from The Cranberry and Its Culture by Benjamin Eastwood, who wrote, "Of this variety there is but one kind. It is about the largest species, and almost as dark colored as blood coral." The Early Black variety, discovered on Nathanial Robbins' Harwich property in 1852, belongs to the bell cranberry family and is now grown on more acres in North America than any other variety.

Left: Seymour's Pond in Harwich, 1989, which Alvin Cahoon drained into nearby Pleasant Lake to make room for his cranberry yards. An abandoned bog (foreground), once belonging to the Cahoon family, now grows wild. (Joseph D. Thomas photograph)

Opposite page: The missionary Richard Bourne, receiving succor of cranberries from the heavenly dove while the powwow weakens. (Pen and ink drawing by Robert A. Henry)

Above: Cyrus Cahoon. Capt. Cahoon was a well-traveled man who went to sea at eleven. As well as a sea captain, he was an auctioneer, a real estate appraiser, a justice of the peace and a lamp black manufacturer. By the 1850s he was widely considered to be an authority on cranberry cultivation. Cyrus developed the Early Black variety, which, so the story goes, his wife Lettice named, remarking, "Ben't they early and ben't they black? Call them Early Blacks." (Ocean Spray Cranberries, Inc.)

Right: Frontispiece to The Cranberry and Its Culture, *by Benjamin Eastwood. Published in 1856, this was the first book on cranberry cultivation, containing a brief history of the young industry, instructions for would-be growers and letters on cultivation from established growers and experimenters. Eastwood was a minister in East Dennis who, under the pen name Septimus, contributed a series of letters on cranberries to the* New York Tribune. *His letters and his book did much to stimulate interest in cranberry cultivation.*

took about seven months to complete the project.

His son Benjamin Cahoon, who was 14 at the time of the project, later wrote:

> Begun in the fall of 1852, the small gang worked diligently and in April of the next year the canal was finished. At the completion on April 1, all the residents of the towns and villages nearby were present. They celebrated the occasion by the ringing of bells and the blowing of horns….
>
> In three weeks, or by the 20th of April, the water in the lake was at least two feet lower than at the first of the month….After this date cranberry bogs were made along the shore of the pond. (From an account by Robert H. Cahoon, in *Cape Cod* magazine, January 1918)

Within a year or so of Alvin's first bogs, his cousin and neighbor at Pleasant Lake, Capt. Cyrus Cahoon, also began developing bogs in Harwich. Both Cahoons were inexperienced, but through trial and error they developed methods of cultivation that gave a foundation to the young industry.

It was at Pleasant Lake that Cyrus Cahoon developed the now famous Early Black variety. Other early cultivated strains named after prominent nineteenth-century Dennis growers are Howes, Smalleys and Sears. The Atkins seedling was named for Joseph Atkins at Pleasant Lake.

In 1854 the first official census of cranberry land took place, an indication of the legitimacy of the crop. In all, there were 197 acres in the Barnstable County towns of Dennis (50), Barnstable (33), Falmouth (26), Provincetown (25), Brewster (21), Harwich (17), Orleans (8), Eastham, Sandwich and Yarmouth (5) and Wellfleet (2). Within ten years, the total cranberry land on Cape Cod was 1074 acres, with Harwich having established itself as a growing leader with 209 acres.

The Rev. Benjamin Eastwood wrote the first book on the cranberry industry in 1856, another indication of its increasing importance. In fact, cranberries may have provided the economic salvation of Cape Cod when, following the Civil War, devastating changes occurred. New metal-hulled ships, steam power and the railroad had a tremendous impact on maritime trade, slowly and

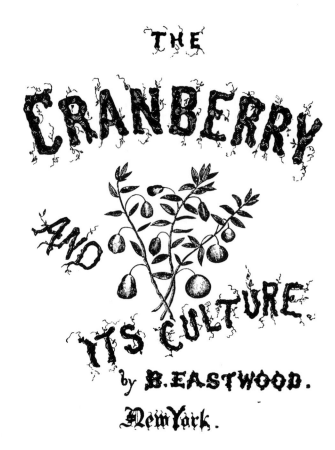

THE CRANBERRY AND ITS CULTURE by B. EASTWOOD. New York.

steadily replacing the wooden sailing vessels that had employed so many Cape Codders. At the same time, the fisheries were experiencing a general depression.

If ever there was a right place and a right time for a new industry, it was Cape Cod in the latter half of the 1800s. Cranberry cultivation seemed ideally suited to local conditions; it utilized previously unusable land; and it offered one of the best yields of any agricultural crop. While thousands rushed to dig for gold in California, Cape Codders were on their knees panning for red gold. Cranberry Fever was under way.

Previously thought to be worthless, low, swampy land was suddenly selling for $50 to $100 an acre. Many of the people buying and developing it were sea captains who had capital and a ready labor force. George E. Thacher had sailing ships in the coastal trade, as well as four or five bogs along the Bass River and others in South Dennis that were flooded by windmills and a tide gate. According to his great-grandson, Linc Thacher of

Dennis, when George's crew stopped sailing in the fall, they went to work on the bogs.

An old poem about a sailor who became a cranberry grower, published in a trade magazine in the 1940s, was attributed to a Captain Bill:

> There's nothing to me in foreign lands
> Like the stuff that grows in Cape Cod Sands;
> There's nothing in sailing of foreign seas
> Equal to getting down on your knees
> And pulling the pizen ivy out;
> I guess I knew what I was about
> When I put by my chart and glass
> And took to growing cranberry sass.

Warner "Dud" Eldredge of Sagamore, now 89, has grown up in the cranberry industry. Both of his grandfathers were cranberry growers: Warren J. Nickerson had bogs in East Harwich, and Josiah Eldredge owned bogs in Chatham and other towns.

Josiah Eldredge had seven boats, a share of seven others and a ship's chandlery. When "business went poor," he didn't shift over from salt to fresh fish, because of his age, but like many others in the maritime industry he turned to cranberries. Dud Eldredge recalls:

> He was quite a cranberry man. He knew how
> to manage bogs and he had a gang of men, so
> people would call him up. I remember my
> grandfather always stopping to talk with
> Elnathan Eldredge down to Chatham. They
> would discuss bogs and roads. My father,
> Charles A. Eldredge, fell heir to the bogs.
> My parents worked on them. Mother kept
> tally. She had ten children, and lost three to
> scarlet fever. When I was young, they took
> me to the bogs with them. I would catch
> frogs to kill time and when I got tired, they
> would lay a cranberry barrel down and put
> me in it with a blanket. When the barrels
> were coopered, they got shaped and the
> staves were tapered. During a certain period,
> they shrank them over a round, iron stove.
> Those barrels would get a little scorched, so
> when I slept in them, it smelt so good.

Cranberry growing had established itself on Cape Cod by the Civil War. It would continue to prosper through-out the rest of the century. The late 1800s and the turn of the century were a watershed for the industry, a break between old and new that saw the rise of Plymouth County as a growing area, the beginnings of the Make-peace empire, the founding of the Cape Cod Cranberry Growers' Association and the establishment of the Cranberry Experiment Station.

Cranberry grower Josiah L. Eldredge, packing cranberries for shipment from Harwich about 1887. Eldredge's daughter Elthea would marry Ellis D. Atwood, Carver cranberry tycoon and founder of Edaville Railroad. At left, on top of a cranberry barrel, note the divided 40-lb. crate box, the earliest standard box used in Massachusetts. (Courtesy of Larry Cole)

"The Indians and English use them much...."

Constance Crosby

Wampanoag legends from Mashpee and Gay Head tell how wild cranberries were a gift from the Great Spirit, brought from heaven in the beak of a dove and dropped into a bog. There they flourished under the care of Granny Squanit, or Squauanit, the traditional women's god. Every fall, when the cranberries were ripe at Gay Head, a young boy was instructed to take a basket of food out into the hollows between the dunes and leave it for Granny Squanit, as a tribute, without looking back. Cranberries may have been a heavenly gift, but it was Granny Squanit, the spiritual guardian of the many wild fruits and herbs the Indians depended on, who ensured a continuing abundance.

The word "cranberry" first appeared in a letter written by the missionary John Eliot describing his second visit to Waban's group of Massachusett Indians one Sunday in the fall of 1647. After preaching the sermon, Eliot called for questions and someone asked "how it comes to passe that the Sea water was salt, and the Land water fresh." He responded rhetorically with another question:

> Why are Strawberries sweet and Cranberries sowre? There is no reason but the wonderfull worke of God that made them so." (*The Day-Breaking*, 1647)

"Craneberries" were again mentioned in 1648, this time among the commodities sold by Massachusett and Wampanoag Indians to the settlers.

In February of 1650, young John Slocume from Taunton set out with a large group "to gather cramberies." He lost his way in the darkening woods and died of exposure; his torn clothes and half-devoured body were found almost a year later. The Wampanoag would have interpreted Slocume's disappearance and death as Granny

Vacciniapaluſtria.
Mariſh Worts.

Squanit's justice. Generally fond of children, especially boys, but angered by the careless way these new white-faced people cut down her forests and took from her gardens without ever acknowledging or thanking her, she led Slocume astray and then changed herself into a beast to revenge herself upon his body.

"It is a delicate Sauce"

In southern New England, Native Americans cultivated maize, beans and squashes for their staples, which they supplemented with many wild fruits, nuts, vegetables, fish, shellfish and game. Wild berries filled an important dietary niche, adding vitamins and flavor to their food. They used berries of all sorts, including cranberries, to make sauces, breads and pudding-like dishes.

Roger Williams observed how the Narragansetts took dried berries, *sautaash*, beat them to a powder and mixed them with parched corn to make *sautauthig*, "a delicate dish…which is as sweet to them as plum or spice cake to the English." (*A Key into the Language*, 1643) They also mashed fresh strawberries in a mortar and mixed them with corn meal to make bread.

Many tribes made pemmican, a mixture of berries, deer, bear or moose fat and dried meat, all pounded together and then formed into cakes and dried for later use. The Chippewa tribe boiled the dried cranberries, *anibimin*, and then seasoned them with maple sugar or combined them with other foods.

When the English arrived in America, there were none of the gooseberry or bayberry bushes they were familiar with waiting to be picked, but there were cranberries. From contacts with local Indian groups the settlers learned how palatable and healthful the native red berries could be. Soon, cranberries replaced gooseberries as a favored filling for tarts or as a sweetened sauce with meat. Mahlon Stacy, an early settler near Trenton, New Jersey, wrote to his brother in England in 1689 that,

> an excellent sauce is made of them for venison, turkeys and other great fowl, and they are better to make tarts than either gooseberries or cherries. (Excerpted in *Cranberries* magazine, 1948)

The Massachusett Indians sold cranberries to the English in the fall and in the spring (spring berries were those left on the vine through the winter). While most berries were consumed fresh, some of the harvest was dried for later use.

Cooking in colonial America was not something done from books. Women relied mostly on family knowledge and recipes learned at home and handed down. In England most fruits were cultivated, while in America cranberries, blueberries, raspberries, blackberries and strawberries grew in wild abundance. Women facing an unfamiliar American environment, with none of the established gardens, herb beds and orchards of Old England, quickly learned to adapt local wild fruits and vegetables to family recipes. Cranberries were added to meat dishes that in England would have been served with a sauce of sweetened "scalded" gooseberries or "Tarte Stuffe"—a barberry conserve.

John Josselyn's 1672 journal, *New-Englands Rareties Discovered*, gave one account of how cranberries were prepared in colonial New England:

> The Indians and English use them much, boyling them with Sugar for Sauce to eat with their Meat; and it is a delicate Sauce, especially for roasted Mutton: Some make tarts with them as with Goose Berries.

The almost "holy" trinity of cranberries, salt cod and corn formed the basis of a distinctive regional cuisine by the end of the seventeenth century. In 1677 the Massachusetts General Court ordered,

> that the Treasurer doe forthwith provide tenn barrells of cranburyes, two hogsheads of special good sampe, and three thousand of cod fish…to be presented to his Majesty, as a present from this Court. (*Records of Massachusetts Bay*, 1854)

The intention was to present King Charles II with the best the colony had to offer in order to appease his wrath over the colony having coined its own money, the pine tree shilling.

The *Boston News Letter*, in 1728, included cranberries as one of the staples of an average family diet, which also included bread, dairy products, meat, root vegetables, raisins, currants, eggs, apples, suet and beer. By mid-century cranberries were prepared and eaten in all sorts of ways. The Swedish botanist, Peter Kalm, while traveling through the northern colonies and Canada, observed how cranberries were,

New-Englands RARITIES Difcovered: IN Birds, Beafts, Fifhes, Serpents, and Plants of that Country. Together with The Phyfical and Chyrurgical REMEDIES wherewith the Natives conftantly ufe to Cure their DISTEMPERS, WOUNDS, and SORES. ALSO A perfect Defcription of an Indian SQUA, in all her Bravery ; with a POEM not improperly conferr'd upon her. LASTLY A CHRONOLOGICAL TABLE of the moft remarkable Paffages in that Country amongft the ENGLISH. Illuftrated with CUTS. By JOHN JOSSELYN, Gent. London, Printed for G. Widdowes at the Green Dragon in St. Pauls Church-yard, 1672.

Title page from New Englands Rarities Discovered, *written in 1672 by John Josselyn, gentleman. Josselyn visited America in 1638 and again in 1663. His 1672 book on the new world's birds, beasts, fishes and serpents describes "cran berries or bear berries" as being like the English night shade, good for "Bruises and dry Blows."*

Opposite page: Sixteenth-century woodcut of the European wild cranberry, called marsh wort *or* fenne berry. *This depiction, probably the earliest known of a cranberry plant, is from A Nievve Herbal, or History of Plants, 1578, by Henry Lyte. In America, wild cranberries abounded in many locations between New England and Virginia. John Josselyn, in New Englands Rarities, described a wild cranberry bog as a salt marsh overgrown with moss, "…the tender Branches (which are reddish) run out in great length, lying flat on the ground where at distances they take root sometimes half a score acres."*

An engraving from a seventeenth-century English cookbook. The cookbooks that women brought with them to the New World bore little resemblance to cookbooks of today or even of the last century. By the mid-1700s many were organized according to the calendar, with recipes using seasonal ingredients given for every month of the year. Even into the early nineteenth century, most English cookbooks concentrated on the preparation of meat, fish, fowl and game and offered no instructions on how to prepare cranberries or many of the other new foods found in the colonies. One sixteenth-century recipe for "Tarte Stuffe" such as barberries instructs, "...take a faire pipkin and fill it full [of barberries]: put in two or three spoonfulls of faire water, then set it upon the hot harth without any coles and so let them boyle while they be soft...and then straine them into a faire bason...let them boyle and put in sugar, and cinamon sufficient to sweeten them and put it into a faire glasse and so use it as you need." (The Good House-wives Treasurie, Edward Allde, 1588)

boiled and prepared in the same manner as we do our red lingon and they are used during winter and part of summer in tarts and other kinds of pastry. But as they are very sour, they require a great deal of sugar. That is not very dear, however, in a country where the sugar plantations are near by. (Benson, *Peter Kalm's Travels*, 1770)

The first cookbook of American foods, *American Cookery*, was published in Hartford, Connecticut in 1796 by Amelia Simmons. She suggested serving roast stuffed turkey or fowl with "cramberry-sauce." Stewed, strained and sweetened cranberries were also baked into tarts.

In 1829 reformer and abolitionist Lydia Maria Child published *The American Frugal Housewife*, aimed at

"those who are not ashamed of economy." The slim volume was filled with recipes for "Cheap Custards," "Common Cooking" and "Hints to persons of moderate fortune." Three of her recipes called for cranberries: cranberry pie, cranberry pudding and cranberry jelly. *The American Frugal Housewife* struck such a responsive chord among American women that by 1832 it was in its 12th printing and went through 32 printings by 1850.

By the mid-1800s, American cranberries became more readily available overseas, and recipes such as "Cranberry and Rice Jelly" began to appear in English cookbooks. While the American cranberry gained popularity, the "other" cranberries were not totally ignored. In nineteenth century England, "fenne-berries," or the European cranberry, became highly sought after for making "marmalade, jelly, jam...puddings and pies" (Eastwood, *The Cranberry and Its Culture*, 1856). The upland cranberry, whose flavor is more astringent than the bog variety, was used for tarts in England and was made into a jelly served with meat in Sweden. As for high-bush cranberries, Peter Kalm thought those growing along the shores of the Saint Lawrence River "had a pleasant acid flavor and tasted right well."

Cranberries as a distinctly American fruit and cranberry sauce as a hallmark of American cuisine were helping to shape an American cultural identity by the early 1800s. To an anonymous Frenchman visiting Boston in 1808, no amount of shared political sentiment or revolutionary brotherhood could shake him of his opinion that Americans were doomed to live in a "most barbarous situation" as long as they ate cranberry sauce, because they would never perceive "the insipidity and staleness of their dishes." He published a short essay entitled "Memoir on the consumption of Cramberry Sauce, by the Americans," in which he claimed that his natural "French enthusiasm" couldn't prevent his being,

> chilled in the daily encounter of huge pieces of half boiled meat, clammy puddings, and ill-concocted hashes, rendered palatable to the natives by a profuse addition of this most villainous sauce. (Tindov, *Miscellanies*, 1821)

Enjoyed by rich and poor alike, cranberry sauce was the great democratizer of American cuisine. New Englanders were especially fond of this "most villainous sauce" and ate it cold at virtually every meal, with fish, fowl, meat and even lobster.

"They are excellent against the Scurvy"

The southern New England tribes called their healing and spiritual specialists, and the accompanying rituals, powwows. The Puritans associated powwowing with devil worship and outlawed it, but in spite of missionary efforts not all of the Wampanoag's traditional spiritual beliefs were abandoned when they became Christians.

Herb lore was considered sacred, dangerous knowledge to be carefully guarded and passed in secret from practitioner to apprentice, usually female. According to Wampanoag herb-doctor Tamson Weeks, diseases were caused by a *tcipai*, the ghost or spirit of someone who had died. When this spirit was removed by means of herbal remedies, the patient was cured. To be effective, these remedies had to be prepared according to certain rules; for example, plants and roots had to be dried in the sun and pounded between two stones or beaten in a special wooden mortar.

Traditional herb lore among early settlers was a combination of European and Native American traditions. In the early 1800s American medical botanists and social reformers became interested in "indigenous medicines" and medical practices. Lydia Maria Child included many remedies from various New England Indian groups in *The American Frugal Housewife*. One in particular recommended a poultice of stewed cranberries to relieve "cancers":

> The Indians have great belief in the efficacy of poultices of stewed cranberries, for the relief of cancers. They apply them fresh and warm every ten or fifteen minutes, night and day. Whether this will effect a cure I know not; I simply know that the Indians strongly recommend it. Salts, or some simple physic, is taken every day during the process.

The American cranberry and the smaller European cranberry were used for a variety of complaints, including blood disorders, stomach ailments, liver problems, scurvy and fevers.

> Fen or Marish Whortes do also quench thirst, and are good against hot fevers, or agues, and against all evil inflammations or heat of blood, and the inward parts. (Lyte, *A Nievve Herbal*, 1619)

[They also] stay vomiting, restore an appetite to meate which was lost by reason of choleriçke and corrupt humors, and are good against the pestilent diseases. (Gerard, *The Herbal*, 1633)

Edward Winslow described how he fed Massasoit "many comfortable conserves, etc., on the point of my knife" which soon revived the ailing Wampanoag sachem, whom many thought was at death's door. (Winslow, *Good News from New England*, 1624)

Cranberry leaves were used for urinary disorders, diarrhea and diabetes. The Shakers at Groton, Massachusetts gathered the leaves of the upland berry from Danvers and sold them, perhaps for their use in the tanning trade as well as for their medicinal properties.

The bearberry and the high-bush cranberry were valued by many Native Americans for food and medicine. The dried leaves of the bearberry, alone or combined with sumac, willow, dogwood or tobacco, made an excellent smoking mixture enjoyed by both Indians and settlers. It was thought that smoking this plant would pro-

An engraving by an unknown artist of Edward Winslow feeding a conserve of what may have been cranberries to Massasoit. Supposedly, the sachem repaid the favor by warning the pilgrims of an Indian attack. John Josselyn wrote that cranberries "...are excellent against the Scurvy. For the heat in Feavers. They are also good to allay the fervour of hot Diseases."

"Theire sitting at meale." Among the earliest artistic renderings of Algonquian life, this watercolor was made by artist/explorer John White at the Virginia colony in 1585. It shows an Indian man and woman eating hominy, made with hulled and boiled kernels of Indian corn. The settlers adopted this meal and ate it with bits of fish and meat. Similarly, the settlers learned of pemmican, a food prepared from lean, dried strips of meat pounded into paste, mixed with fat and berries and pressed into small cakes. Pemmican had an excellent keeping quality and was made primarily for taking on journeys.

Below: An engraving of the bugle cranberry, from Eastwood. The bugle was named not for the musical instrument but for the bugle bead, a women's fashion accessory of the nineteenth century. This variety was noted for its excellent keeping quality, though, according to Eastwood, it didn't color as well as the bell variety.

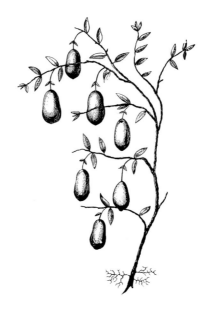

tect against malaria, and Indians made a lotion from the leaves to treat venereal disease. Both Indians and settlers made a tea from the dried leaves to act as a diuretic and to cleanse the urinary tract. The Penobscot and Malecite Indians used the high-bush cranberry to treat swollen glands and the mumps.

Scurvy, resulting from a lack of fresh foods and especially vitamin C, was a serious problem for many of the early explorers and colonizers who wintered in New England or Canada. They were often ignorant of nearby plants, such as cranberries, that could have saved them much suffering and death during the cold months. Soon, however, the colonists learned that these "very wholesome fruits which may be kept till fruit come in again" would keep them in good health (Mahlon Stacy, 1689).

Just as British ships were known for their limes, American vessels became famous for their cranberries. By the 1800s, cranberries were regularly loaded by the barrel aboard whalers and clipper ships to provide fresh fruit and vitamins during long voyages. In Herman Melville's *Moby Dick*, one whaleman derided his captain over the thought of a voyage without cranberries:

> Go out with that crazy Captain Ahab?
> Never! He flat refused to take cranberries
> aboard. A man could get scurvy, or worse,
> whaling with the likes of 'im.

"In the Autumn they sell Craneberries"
(Shepard, *The Cleare Sun-Shine of the Gospel*, 1648)

The Indians harvested cranberries both for their own use and to trade or sell to settlers. By the late 1640s, converted Massachuset Indians had established "Craneberries" as one of many items, including "Brooms, Staves, Ele pots, Baskets, Turkies, Fish, Strawberries, Venison, etc.," that they harvested, hunted or manufactured to sell to the English.

Wild cranberries growing on the common lands in the

Cranberry Day on Gay Head, Marthas Vineyard, 1930s. At the front of the cart is James Cooper, the Wampanoag cranberry agent from 1912 to 1923; at the far right is Frank Manning, who owned and harvested a wild cranberry bog in Lobsterville; and in the wagon is Gracie Manning.

The wild cranberry crop dapples the rugged Vineyard landscape from Chappaquiddick to Cedar Tree Neck. Bogs flourish near the heads of the Vineyard's Great Ponds, including Jane's Cove in Edgartown and Long Cove in Tisbury. In the 1930s the Wampanoags sorted through nearly 40 acres of windswept bogs in the Lobsterville peninsula, but hurricanes in 1938 and 1944 eradicated the bulk of the crop. Before that major damage, the harvest was used in the home for sauces and baking and also bartered for groceries down-Island or traded for goods like molasses in New Bedford.

Gladys Widdiss, whose Indian name means "wild cranberry," remembers earlier cranberry days. "When I was a girl, before automobile time, when we lived on the Manning homestead off Lobsterville Road, we got up before sunrise. Cranberry Day really began before Cranberry Day. Mothers started cooking for the lunch two to three days before. It was like a great big picnic. Each family had a specialty—something they liked to cook. Leander Smalley made mincemeat pies from scratch. We walked and rode in the oxcart my grandfather Thomas Manning owned."

The town of Gay Head, which before incorporation in 1870 had been an Indian reservation, annually appoints a cranberry agent who offically opens the harvest. This custom has deep roots in the past, when the medicine man informed the tribe that the berries were ready for harvest and performed the magical rites over the dunes in acknowledgment that the cranberry was a gift of the great spirit, or Manitou.

An 1845 state law protected the Gay Head Indians' cranberry crop from "their thoughtless White neighbors," but many Whites had their own bogs. At the turn of the century, the Daggetts of Cedar Tree Neck harvested and sold berries from wild bogs on their homestead. And in 1882 Capt. James L. Smith of Vineyard Haven shipped hundreds of barrels of cranberries to New York, probably from old Uncle Smith's bog, which was likely the first wild bog to be improved and developed.

Vineyard Gazette accounts from 1882 state that school in Vineyard Haven closed the first week in October so that pupils could help Smith pick his berries, just as today schools in West Tisbury and Tisbury excuse Gay Head Indian children so that they can take part in Cranberry Day. (Photograph from Ocean Spray Cranberries, Inc.; caption notes by Elaine Lembo)

early towns were at first available to all. However, as demand increased and prices rose in the early 1800s, communities such as Barnstable and Gay Head passed laws to reserve the common land cranberry harvest for their townspeople or tribe. Barnstable's Sandy Neck abounded in wild cranberries, which area residents, including the Mashpee Indians, had enjoyed for many years. By the early 1820s, picking of unripe berries and picking by outsiders became such a problem that the town sought the state's help to control the situation. The state refused to intervene, so selectmen took it upon themselves to set the day for first harvest and fine anyone who picked before that date or who was not a town resident. These measures may have been aimed at restricting Mashpee's access to Sandy Neck. A certain percentage of the berries picked belonged to the town, which it sold to raise revenue.

In 1842 members of the Gay Head tribe petitioned the state legislature for a fine to be imposed,

> on such white persons as shall gather cranberries on the land of the Indians, without their consent…[because] some of our

thoughtless White Neighbours have been and still appear to be desirous of taking from us our means of a living and supporting our poor. (Massachusetts State Archives)

The situation had grown so serious that the Gay Headers were "under the Necesity of having a Watch kept while the Berries were a ripening." The petitioners described how the "Considerable quantities of Cranberries," growing on lands "handed down to us by our Fathers," annually raised $100 to $300,

> which principally falls into the hands of the most Indigent of the Women and Children of our tribe who gather the Most of the Berries and which to them is a Staple means of support through the winter.

This time the legislature listened and in 1845 passed "An Act for the Protection of Cranberries on Gay Head" (Acts and Resolves, 1845, Chpt. 202). The Indian proprietors of Gay Head were empowered to set the date for the annual harvest and fine any of their tribe who

picked before that date and also to fine anyone not a proprietor, or acting on one's behalf, who picked cranberries before the tenth day after the opening of the harvest.

Wild cranberries were harvested from privately owned marshes and meadows, and it was not until demand began to outstrip the wild supply that owners of cranberry lands became concerned with who picked their berries. Picking on these lands was sometimes done by halves, with equal portions going to the picker and the owner. Even when cash was paid for picking, the rates were low. In 1830 Henry Hall, the pioneer of cranberry culture, paid 20 cents per bushel. Considering that pickers managed three to four bushels daily, that meant they earned 60 to 80 cents.

The earliest mention of cranberries being sold is the observation made by John Eliot in 1647 that Massachusett Indian converts sold cranberries to the English in the spring and fall. Others mentioned receiving cranberries from local Indians. Mahlon Stacy, in 1689, said, "We have them brought to our homes by the Indians in great plenty," and near the Columbia River in Oregon, Lewis and Clark paid "high prices" to a group of Chinook women for "Cramberies," salmon, small baskets and mats. (*The Journals*, 1805)

As wild cranberries became more generally known and prized as a foodstuff throughout the colonies, not everyone could simply pick or buy locally. Swedish botanist Peter Kalm, who traveled throughout the northeast in 1750, noted that cranberries were brought to market every Wednesday and Saturday at Philadelphia late in autumn. Some of these were then preserved and shipped to the West Indies and to Europe. In London American cranberries had to compete with those imported from czarist Russia in "certain quaint-looking earthen jars." The outbreak of the Crimean War in 1853 halted the flow of "these Muscovite luxuries" to London, thus giving the American cranberry a chance to become established in the English market (Eastwood, *The Cranberry and Its Culture*, 1856).

In the early nineteenth century, the desire for more abundant cranberry harvests and a dependable cash crop became linked to an interest in agricultural innovation. Landowners were urged to turn what were considered "useless" swamps, bogs, wet meadows and marshes into something productive and profitable, even though many of these areas were already producing hay, pasture for grazing animals and bog iron for blast furnaces and foundries. Throughout the next century these same meadows became the focus of great experimentation and innovation, but this time for the berries they produced.

"Why are Strawberries sweet and Cranberries sowre...."

To the Christian Mashpee Indians of the seventeenth- and eighteenth-centuries, the story of Richard Bourne's triumph over one of their powwows may have told of more than the origin of the cranberry. There is traditional mythology in the dove bringing the first cranberry to earth in its beak: The seventeenth-century Indians in southern New England believed that a crow brought the first seeds of Indian corn and beans from the creator Cautantowit's field in the southwest; Bourne, who from all accounts was a sort of culture hero to the Mashpee, met his Indian opponent in a test of spiritual strength that was traditional among rival Indian powwows; and the dove, a potent christian symbol, coming to Bourne's aid was actually his spirit helper—every powwow had at least one such spirit helper to call upon in times of need.

By symbolically placing Bourne in the role of a traditional powwow, the Mashpee were transforming Christianity into something more compatible with their own

religious traditions and beliefs regarding spiritual power. At the same time they were negating the Puritan interpretation that once again Indian "devil worship" had been overcome.

While the English and Indian might acknowledge that cranberries were God's creation and gift, the similarities in their world views ended there. To the English colonists in Massachusetts, the sale of cranberries and other items by the recently converted Indians was a sign of their growing industriousness and participation in the market economy. To the Indians, cranberries were part of a material and spiritual world that was balanced according to reciprocal relations between all things. The continued abundance of wild cranberries, game, fish, nuts, fruits, herbs and roots depended upon the Indian's continued respect for Granny Squanit. Her anger and mischief reminded people that they had obligations to care for each other and for the land where they lived.

Top and Bottom: Cranberry Day, Gay Head, 1989. Held on the second Tuesday of October, Cranberry Day is when the Gay Head Wampanoags harvest the wild meadows near the cliffs. The meadows and picking privileges belong exclusively to all Gay Head Wampanoags. In 1987 the 238 acres of Gay Head common lands, including the cranberry bogs of Lobsterville, were marked for return from the town to the tribe in its historic land claims settlement after a bitter 14-year struggle. That year the U.S. government granted the Wampanoags tribal status and Congress passed legislation conveying to the Indians the cranberry bogs, the herring creek and the face of the clay cliffs on the promontory edge.

Gladys Widdiss, former tribal council president, says of Cranberry Day: "With the recognition of the tribe, there's going to be more incentive to focus on Cranberry Day. Personally, I hope they make it an official homecoming day for Wampanoags, so that they will be asked to come home for the day and reinforce their identity as tribal members."

In more modern times, Eben Bodfish of Oak Bluffs harvested berries from centuries-old wild bogs off Lambert's Cove Road and in West Tisbury. The Vineyard Gazette reported in the fall of 1935 that he was harvesting "cranberries the size of horse chestnuts." Manual Duarte of Tisbury later bought Bodfish's Lambert's Cove bog and in 1954 shipped 108 barrels of Early Black and McFarlin berries to the Ocean Spray plant in Onset.

Duarte's cranberry operation hasn't been active since 1970, but in 1981 his six acres of bog were bought by the Vineyard Open Land Foundation (VOLF), which hopes to find someone to lease the bogs and get them going again. VOLF believes working bogs would be a good use of open space as well as provide income for its land conservation activities. (Caption notes by Elaine Lembo; photographs by Joseph D. Thomas)

The Early Cultivators

Robert Demanche

Nobscusset Point, Dennis, Cape Cod. To the north, Cape Cod Bay. To the west, Sandy Neck. To the east, Sesuit Point, and beyond, Quivet Neck—all bordering the cold bay waters that shape the sandy shoreline. To the south sits Scargo Lake, and then Scargo Hill, the highest spot in Barnstable County and the first point of land that sailors spied on their return home.

At the North Dennis fishing village of Nobscusset, at water's edge, the dunes stood as a barrier to the low wetlands behind them. Further inland, pitch pine and white, red and black oak sparsely covered the terrain. Around 1816, fifty-two dwellings, the old meeting house, a Freemason lodge and two windmills populated the village, and near it stood strings of saltworks and their windmills along the bay. It was here, on land near the shore, that Henry Hall first cultivated cranberries.

Cape Cod Pioneers

Henry Hall's success with his cranberry yards did not go unnoticed by his Dennis neighbors and relatives. Many of them followed suit, building their early bogs close to the seashore and even diking off and planting on salt meadows that had been freshened by rain and snow. It was simpler this way since there were no trees or brush to clear on salt meadows, and beach sand was nearby.

Elkanah Sears and his son William set out vines at Scargo Lake in 1819 and became profitable growers. Sears, like Henry Hall, was both a Revolutionary War veteran and a saltmaker. By the 1840s, Asa Shiverick, the patriarch of a Dennis shipbuilding family and an important saltmaker, also decided to take up cranberry growing. He left his shipyard to his sons, who built the only clipper ships to come from Cape Cod. Other growers from Dennis were Aaron and Ebeneezer Crowell and James Anthony Smalley, who began in Dennis in 1853 and developed the Smalley Howe variety. Eli Howes

began in 1843; his son James Paine Howes developed the Howes variety.

Early growers found local newspapers invaluable for disseminating cranberry news. Between 1820 and 1840, the Falmouth *Nautical Intelligencer*, the Barnstable *Journal*, the Yarmouth *Register* and the Barnstable *Patriot* all carried accounts of cranberry cultivation. Growers also read about the fledgling industry in journals such as *The Naturalist*, the *New England Farmer*, *The Massachusetts Ploughman*, *The Orchardist* and *The Cultivator*.

During the nineteenth century county agricultural societies throughout Massachusetts held annual fairs and awarded cash premiums for the largest, best-looking, best-tasting fruit and vegetables. Since entrants in these competitions had to submit an explanation of their growing methods, all growers shared in the winners' success. By 1850 at least seven agricultural societies—Plymouth, Barnstable, Bristol, Norfolk, Essex, Hampshire and Franklin—awarded premiums for cranberries.

Dennis remained the heart of cranberry growing until the mid-1850s, but the industry was also taking hold in other parts of Barnstable County. James Lovell of Osterville and Leonard Lumbert of Centerville were two early growers who began cultivating in the 1830s. Lovell achieved mixed results against cranberry worm in the mid-1840s by applying wood ashes, fine salt and lime as "pesticides," and he won the first premium ever offered by the Barnstable County Agricultural Society. Lumbert protected his crop from frost by covering the vines with

cotton cloth placed two feet above the bog surface.

In Sandwich Solomon Hoxie began growing cranberries by 1846. Hoxie, who at age 18 made foreign voyages on merchant ships, provided a glimpse of the industry in his "Cranberry Business" account book, in which he recorded his labor costs, selling prices and experiments. He referred to sanding and harvesting, planting in the spring and fall and planting at various locations: "below the fence," "next to the dike," "below the orchard," "southeast of the grapevine" and at the "east end of the pond." He didn't record his results, but we know that by 1850 his bogs paid off:

> Sold Ala Hallway 4 bushels of cranberries, $10.35...Sold David Akin and others at Bass River, one and three-quarters bushels cranberries, $4.38.

Hoxie also recorded his labor expenses, crediting Philip H. Robinson with "one day's work at 60 cents" in 1846 and "E. Jones for six days of vine setting, clearing

Left: Hiram Hall, Henry's son and heir to his cranberry yards. After taking over from his father in 1834, Hiram became one of the Cape's most successful growers and was a founding member of the Cape Cod Cranberry Growers' Association in 1888. (Ocean Spray Cranberries, Inc.)

Below: Henry Hall's first bog at Nobscusset. There has been some debate about the actual location of the first cultivated bog on the Cape. Some say it was near Hall's saltworks; others say it would have been near his house by Hiram Pond, which also had swampy shores and wind-blown sand. Cranberry historian Clarence Hall believes it was the one pictured here, a four-acre bog behind Henry's small cottage near the shore. (Ocean Spray Cranberries, Inc.)

Opposite page: Title unknown, oil on canvas, by Eastman Johnson, circa 1872. Johnson, who had a summer house on Nantucket, gave his island cranberry harvest paintings a lyrical, impressionistic quality that did not always match the reality of his subject. (Massachusetts Cranberry Experiment Station.)

and such work" at 75 cents per day in 1853. He was not above hiring himself out as a worker. In 1849 he wrote, "I set out vines for Francis Jones for one day," and in 1869, at age 61, he hoed Edward Wing's bog for eight hours for $1.20.

Braley Jenkins began cultivating around 1852 on Sandy Neck, and by 1861 he had acquired 50 acres of bog as well as a reputation for quality berries. The lack of practical roads on Sandy Neck made transporting the crop and traveling back and forth difficult, so Jenkins built a large house at his bog for his 25 or so workers and brought his berries to the mainland on his schooner, the *Pomona*, which carried as many as 40 barrels per trip. Sometimes he shipped part of his crop long distances in hogsheads filled with water.

In Brewster of 102 people taxed in 1860 for any holding, 42 were assessed for their cranberry properties. The previous year, Brewster growers had raised 90 barrels of cranberries worth $3848.

Josiah Freeman, originally a saltmaker, was the first to be taxed for bog in Orleans. He shipped his cranberries and salt to market on the *Bay Queen*, the last of the Orleans packets, and later became one of the first directors of the Cape Cod Cranberry Growers' Association. Freeman is remembered for his erroneous notion that the cranberry was a marine plant. He preferred to plant the vines as near to the shore as possible without having the sea water flow over them.

Russell Hinckley of Marstons Mills was not satisfied with vines from his own area, preferring those from Sandy Neck. He covered his quarter-acre pasture with six inches of beach sand to "kill out the grass" and enclosed the area with a ditch 3 feet wide and 3 feet deep "to answer the purpose of a fence to keep out straying cattle." He harvested 12 bushels in 1849 and took first prize for his fruit at the Barnstable County Fair.

Edward Thacher was possibly Yarmouth's first grower, having prepared his bogs in 1846 on Weir Road, not far from Upper and Follen's ponds. He planted on a former mill pond, scalped some adjoining upland and had the foresight to transplant vines he had marked the year before when they were bearing good fruit. He won first prize at the fair twice.

In Eastham the Board of Assessors began listing many peat bogs and meadows as cranberry swamps by 1858. At least 15 were taxed as such, belonging to Joshua Cole, probably Eastham's leading early cranberry man; Knowles and Roland Doane; John, Joshua, Josiah and

Asa Shiverick (b. 1790), sea captain, saltmaker, cranberry grower. In 1855 he was the number-one grower in Dennis, having produced 117 barrels of cranberries worth $1404. Asa came from a devout family, said to start the sabbath at sundown on Saturday and end it at dawn on Monday. (Ocean Spray Cranberries, Inc.)

Right: Jay Bird Bog stencil. The shipper's brand was applied to the head of the barrel by brushing shoe polish over the stencil. Jay Bird Bog belonged to Braley Jenkins. (Courtesy of Jim Jenkins)

Edith Higgins; Ruben Nickerson; Joshua Paine; James Rogers; Louis and Nathan Smith; Amos and Jonathan Sherman; and William Wareham.

Cranberry cultivation in Provincetown began in the 1840s and flourished for another 20 or 30 years. At first, residents resented the fact that individuals and companies obtained leases to build bogs on land that had formerly been open to all to gather the wild berries. Thomas Lothrop began cultivating in Provincetown about 1847. In 1855 he proposed to build a 70-acre bog at Shank Painter Pond, at a time when such large-scale thinking was unheard of. Lothrop never completely realized his goal, but he did arouse a great interest in cranberry cultivation on the lower Cape.

In Lothrop's time, bogs filled almost every hollow between the dunes in Provincetown, but not a single acre of cultivated bog remained by 1950. With no brooks or streams to dam and unstable, shifting sand dunes, cranberry growing gave way to more lucrative livelihoods.

Harwich Takes the Lead

Beginning in the early 1830s, Harwich growers took up cranberry growing with a passion. Many were former sea captains: Alvin Cahoon, Cyrus Cahoon, Zebina Small, Nathaniel and Abiathar Doane and Nathaniel Robbins. It was Cyrus Cahoon who in the mid-1830s revolutionized cranberry cultivation with the first level-floored cranberry bog. By 1859 Harwich had overtaken Dennis in cranberry acreage.

Obedia Brooks Jr. of Harwich prepared a marketing study of the industry in 1859 that listed all the growers in several Cape towns. Published in the Yarmouth *Register*, the study showed that Harwich, with a yearly total of 856 barrels valued at $10,137, had overtaken Dennis, which had 831 barrels valued at $9637. No less than 151

growers were listed, a jump from the "21 plus a few Sundry Persons," listed four years earlier. Since 1855 the total Cape crop had risen from 856 barrels to 2597, and its value had increased from $10,137 to $26,934.

The top five Harwich growers were:

	Barrels	Price	Total value
Albert Clark	269	$13.50	$3520
Susanna Winslow	207	11.00	2343
Z.H. Small	108	11.00	1188
Nathan'l Winslow	53	13.00	689
Alvin Cahoon	45	13.00	455

The top Dennis growers were:

Asa Shiverick	117	$12.00	$1404
P.S. Crowell	84	12.00	972
Jonathan Tobey	50	12.00	600
Philip Vincent	46	11.00	506
Hiram Hall	45	11.00	495

Cape Cod was not alone in its quest for successful cranberry cultivation. Growers in Middlesex, Norfolk, Essex and Plymouth counties were also experimenting with various cultivation techniques. For example, methods in flooding and sanding were pioneered by a Middlesex County grower, Augustus H. Leland of Sherborn, who began cultivating in 1838. Leland flooded his bog to kill the "cranberry worm" and protect his vines against frost. He "ice-sanded" by spreading sand, soil and gravel onto his bog when the flood had frozen.

Deacon Addison Flint of North Reading favored different cultural methods. Beginning in 1843, he cleared his swamps by flooding them for three years to kill all the grasses, bushes and moss and then burned what remained. He claimed to plant successfully with refuse berries from the previous year's crop by "crushing each berry between the thumb and finger and placing it just under the sod."

Sullivan Bates from Bellingham in Norfolk County was an advocate of highland or "upland" cultivation. His method was to plow the land, spread on a quantity of "swamp muck" and then, after harrowing, set out the plants in furrows 20 inches apart. He didn't flood the plants or use sand, yet he claimed to gather as many as 200 bushels per acre and grew his fruit double the size of swamp-grown berries. Sullivan's methods were widely discussed by growers throughout the state.

Capt. Winthrop Low, Essex County's first cultivator, won the 1847 county agricultural society premium of $15 for his upland-grown berries, as well as the acclaim of a society representative:

> The soil cultivated by him was, most of it, perfect English corn land and [as] an example of its [upland] nature, a row of white beans was planted between each two rows of cranberries.

Left: Obedia Brooks Jr. (b. 1809). Brooks had a background in business and commerce, but he was also a cranberry grower, as well as chairman of the cranberry committee of the Barnstable County Agricultural Society. His 1859 study of cranberry growing on Cape Cod in the Yarmouth Register "[threw] a good deal of light upon the increase of this branch of home industry." (Ocean Spray Cranberries, Inc.)

Below: Title unknown, oil on canvas, by Eastman Johnson, ca. 1872, showing a lone cranberry harvester near the dunes on Nantucket. (Massachusetts Cranberry Experiment Station)

Yankee soldiers, dining on a Thanksgiving meal of turkey and cranberries, 1864. At this point in the war Sherman had taken Atlanta and time was running out for the Confederacy. Dr. J. Marcus Rice, a surgeon with 25th Regiment of Massachusetts Volunteers, was a prisoner of the Confederates at Libby Prison in Richmond, Virginia. In an 1863 letter to "my dear wife" he enclosed a list of articles he wished her to send him. Third on the list was cranberries. (Pen and ink drawing by Robert A. Henry)

In spite of such testimonies, Benjamin Eastwood, in *The Cranberry and its Culture*, warned against upland cultivation.

By 1855 Middlesex County, with 2554 acres, Worcester County, with 641, and Norfolk County, with 897, boasted many times the acreage of Barnstable's paltry 197. A likely explanation for this imbalance is that most of the off-Cape acreage was wild cranberry patches rather than man-made, cultivated bogs. Cranberry historian Josh Hall claimed that the reason the cranberry industry succeeded in Barnstable County but failed elsewhere was that Cape growers heeded Eastwood's warning and built their bogs in sandy swampland rather than on the drier uplands.

Cranberry Fever

By the late 1850s as cranberry growing spread throughout eastern Massachusetts, agriculturists urged landowners to convert their worthless swamps into productive cranberry yards. "Cranberry Fever" was taking hold, spurred on by a new generation of growers on Cape Cod and in Plymouth County who could now make their living from cranberries. The Barnstable *Patriot* declared in September 1855 that growers "never received such

enormous profits as this fall."

In 1854 Massachusetts established its State Board of Agriculture, and the following year the secretary of agriculture, Charles L. Flint, visited more than 100 cranberry "plantations" throughout the state. He noted that Cape Cod grew the finest cranberries, particularly "...a large, round, and black cranberry...it might properly be called the Black cranberry."

Cranberry fever, and the prosperity it brought, continued throughout the 1850s and '60s. The Civil War years proved particularly beneficial for the industry. In April 1861 Joseph L. Daniels wrote in the *New England Farmer* that cranberries were never in such demand, selling for $10 to $20 a barrel and "never less than $8.00." According to Clarence Hall,

> Everywhere in the cranberry regions recruits were falling into the ranks of the industry. New communities were swelling the resources of the cranberry army. The War of the Rebellion was not to dampen cranberry enthusiasm, but to continue the boom, with high prices for cranberry land and crops...

In Harwich alone, the 1862 harvest of 1500 barrels was nearly double that of 1855, and in some isolated cases a single barrel of cranberries fetched up to $50.

For Thanksgiving in 1864, General Grant's soldiers dined on 75 tons of poultry, several hundred barrels of apples, canned fruit, pickles and cranberries, shipped to them from New York. Joseph J. White wrote,

> A cheap article—canned cranberries—was manufactured during the war by using half a pound of brown sugar to each quart of cranberries. (*Cranberry Culture*, 1870)

In 1863 the Lincoln administration created the U.S. Department of Agriculture, and the Commonwealth established the Massachusetts Agricultural College, forerunner to the University of Massachusetts. Also in that year, the great Swiss naturalist Louis Agassiz declared at the Barnstable Fair that the geologic formation of the Cape favored cranberry cultivation. Barnabas Thacher of Yarmouth, then living in Boston, is believed to have tried to develop a cranberry picking machine in 1863, and for the first time, cranberry prices were reported by the American Cranberry Growers' Association.

By the time the war ended in 1865, total acreage in Barnstable County had increased to five times what it had been ten years earlier.

Early Plymouth Planters

Among the first growers to arrive in Plymouth County were Calvin, Hiram and Hiram E. Crowell of Sandwich. In 1856 they built a bog at White Island Pond, near the boundaries of Wareham, Bourne and Plymouth, that later became LeBaron R. Barker's well-known Century Bog. Benjamin D. Phinney's bog, built in 1856 in the "Darby" area of North Carver, was probably the first cultivated bog in that town.

Wild berries were once harvested in Carver from "patches" in Carver's soggy, natural-hay meadows. The most famous of these was a 500-acre, partly wooded swamp-meadow called "New Found Meadows," or "New Meadows," which later became part of the Ellis D. Atwood company bogs. In early times, peddlers traded their wares for cranberries at New Meadows, along Old Rochester Road, and shipped the fruit by stagecoach to markets in Boston and New Bedford.

In 1876 George P. Bowers, a prominent iron industry man, interested Cape grower Abel D. Makepeace in growing cranberries in Carver. He sold Makepeace 38 acres, which became known as the "Carver Bog." In 1878, Makepeace helped Bowers build the 25-acre East Head Bog. The success of the East Head Bog and later Makepeace's Wankinco Bog launched Plymouth County as the new center of the industry.

In 1885 Plymouth County had 1347 acres under cultivation and Barnstable County had 2408. By 1895 Plymouth had overtaken Barnstable, 3766 to 3255. Plymouth produced about 48 percent of the state crop by 1905 and 61 percent by 1915. Carver alone tripled its cranberry acreage between 1890 and 1912.

The use of swamp land in Carver was changing for the third time in as many centuries. Historian Henry S. Griffith wrote that Carver swamps,

> furnished the residents of this region with pasturage and hay during their first century, with bog ore for the operation of their furnaces during the second century, [and in the next] proved to be ideal ground for the cultivation of cranberries...

George Bartholomew wrote about the Carver iron industry in 1935, describing the rich iron ore deposits in many of Carver's ponds, bogs and swamps. He told of the beginnings of the iron industry there:

> ...Ore was extracted, blast furnaces were set up, wood from the adjacent forest was converted into charcoal, [and] sea shells were obtained from nearby Cape Cod and Buzzards Bay. All through most of the eighteenth and some of the early nineteenth centuries, native ore was converted into household and farm utensils, and some of those foundries continued to operate for many years after the native ore had been exhausted....
>
> ...On those same bogs and marshes and along the borders of some of those same ponds where those iron men of Carver extracted ore, grew the wild cranberry that has become a part of the nation's menu. In my boyhood days when the iron industry of Carver was fading away, some of the old iron masters that I knew turned their attention to the cultivation of cranberries, and Carver has since become one of the greatest cranberry centers of the world.

Handpicking on a Carver bog, ca. 1890s. Once large-scale cultivation came to Carver, cranberries were no longer considered common property and harvesting became a transaction between owner and picker. An early method of payment was "by the halves," the owner keeping one half of a picker's berries and the picker keeping the other half. (Ocean Spray Cranberries, Inc.)

Beyond Cape Cod

Robert Demanche

The Jersey Pines

English settlers in New Jersey were well acquainted with the cranberry. As early as 1789, the New Jersey legislature passed a law protecting the wild, ripening fruit and levied a ten-shilling penalty for picking before October 10.

Around the mid-1830s, Benjamin Thomas started a small cranberry patch on the edge of Burr's mill pond near Pemberton. William Braddock first planted near Medford in 1848. In the fifties and sixties, Joseph C. Hinchsman, Daniel H. Shreve and Theodore Budd, three of New Jersey's major growers, started their bogs.

John I. "Peg-Leg" Webb of Cassville, Ocean County, was probably New Jersey's best-known early grower. He was called Peg-Leg because as a boy he lost a leg to a falling tree. (Local folklore has it that a hot cranberry poultice was applied to the injury in an attempt to save the leg.) Webb became a successful grower by experimenting with different ways of cultivating wild vines. He is said to have used his peg-leg as a dibble to press vines into the soil, and he was the first in New Jersey to discover the benefits of covering the bog muck with sand before planting. According to the *New York Times*, this was "…a discovery that subsequently filled John Webb's coffers with bright round dollars."

Peg-Leg stored his berries on the second floor of his barn, and because of his handicap he retrieved them by pouring them down the stairs. One day he noticed that the sound berries bounced down to the bottom while the poor ones, which didn't bounce, stayed on the steps. He had discovered the bounce principle, which has been used since in various cranberry sorting processes and machines to separate good and bad berries.

The cranberry fever that gripped Massachusetts in the 1850s also hit New Jersey. By 1870 the state moved ahead of Massachusetts as the nation's largest producer of berries, even though this expansion was not to last. Unlike the Cape Codders, New Jersey growers paid less attention to sound bog building. Bogs might be four or five feet out of level, tree trunks were not removed, little weeding or sanding was done and adequate water supplies were not maintained.

New Jersey growers put together the first cranberry growers' association in the country in 1864. Theodore Budd called the first meeting in Vincentown by announcing: "Boys, let us get together and compare notes. Let us mark these danger spots or blast them out of existence." A second group, the New Jersey Cranberry Association, was formed several years later. The two groups joined in 1873 and became the American Cranberry Growers' Association.

Even as late as the mid-1880s, New Jersey growers planted native berries, marketing them as "Jerseys." Later they imported vines from Massachusetts and Wisconsin. Many New Jersey berries found a market at Charles Wilkinson's produce house in Philadelphia, established in 1861. By the 1870s, Philadelphia overtook Boston and New York as the leading cranberry market in the country. Wilkinson & Sons was the city's first cran-

cranberry commission merchant. One son, Charles W. Wilkinson, later helped organize the Growers' Cranberry Company in 1895 and became general sales counsel for the American Cranberry Exchange in 1911.

James Fenwick, who patented one of the earliest versions of a cranberry separator, set out the first vines of what would become Whitesbog in the early 1860s. One of the largest cranberry bogs in the country, Whitesbog was completed by Joseph J. White, Fenwick's son-in-law and author of the 1870 book, *Cranberry Culture*. Joseph's daughter Elizabeth ran Whitesbog, which had its own post office, general store, housing compound and power plant, throughout the mid-twentieth century. She helped advance cranberry and blueberry cultivation, continued expanding Whitesbog and provided leadership in the New Jersey Growers' Association.

From the early years of the twentieth century until the mid-1930s, New Jersey berries accounted for about 40 percent of the U.S. crop. But drought, disease, insect problems and the Depression sent the industry into a downspin from which it recovered only in the mid-1960s. Growers have since increased their yield per acre and now produce about 10 percent of the national crop.

Cranberries have always been grown chiefly in Ocean, Burlington and Atlantic counties, which make up much of New Jersey's "Pine Barrens." This is a large, isolated region in the central part of the state whose scrub pine,

sandy soil and vast wetlands are much like the Cape and Plymouth County growing regions of Massachusetts, particularly Carver. Like Carver, the Pine Barrens were once an important iron-producing region.

The Wisconsin Marshes

As early as 1700, Native Americans in Wisconsin gathered what they called "atoqua," or cranberries. The Indians and the early white settlers harvested the wild fruit for their own use and to trade and sell, just as their counterparts did in Massachusetts. In 1853 a Mr. Peffer became probably the first to cultivate the cranberry in Wisconsin, but it was Edward Sackett of Aurora, near Berlin, who in the early 1860s became the state's first large-scale grower.

Early growers merely fenced off naturally occurring vines and cut down trees and brush in the marshlands where they grew. These marshes held bogs of soil sur-

An engraving of a peck box from Cranberry Culture by Joseph J. White, 1870. The peck box—13 by 8 inches square and 6 inches deep, with a handle—was introduced in New Jersey by 1870 to replace the more cumbersome barrels into which pickers dumped their full picking containers. A peck was not always a peck to growers. Many used boxes that held two to three more quarts than the standard peck "...to allow for waste" such as vines and weeds, according to one New Jersey grower in a 1911 study of cranberry labor. However, investigators noted that tallykeepers on New Jersey bogs required pickers to remove waste before their boxes could be checked.

Left: Though the Whites stood out for their enlightened attitude toward their workers, by most accounts, working conditions on the Jersey bogs were harsh, particularly for children, in the first quarter of the twentieth century. By its failure to pass any child labor legislation, the state of New Jersey condoned this exploitation. Photographer Lewis Hine took this picture of 8-year-old Jennie Camillo at Theodore Budd's bog at Turkeytown, near Pemberton in 1910. (Ocean Spray Cranberries, Inc.)

Far Left: Elizabeth White, circa 1915. Although her lineage in the New Jersey cranberry industry was impeccable—granddaughter of both James Fenwick and Barclay White, pioneer New Jersey growers, and daughter of Joseph J. White, author of Cranberry Culture—Elizabeth White did not rest on these credentials. She did much important work in blueberry and cranberry cultivation; she served as vice president of Joseph J. White, Inc., parent company of Whitesbog; and, like her father, she was well known as a philanthropist and social reformer. (Ocean Spray Cranberries, Inc.)

rounded by shallow water or muck and covered with grasses and small brush. Growers soon found that ditching and draining these bogs increased their production.

Berlin County was the early focus of cranberry growing in Wisconsin, but it gradually gave way to the swamplands of glacial Lake Wisconsin, now known as the Cranmoor and Mather-Warrens districts, in the central part of the state. In the 1880s the industry spread into Juneau, Jackson and Monroe Counties and later into northwestern and north-central Wisconsin. Now it is centered in Wood, Jackson and Juneau Counties.

In 1905 growers found that they could gather twice as many berries if they flooded their marshes to the tops of the vines first and then scooped or "raked" the floating berries. They later devised mechanical water-pickers such as the Case and the Getsinger. Water-harvesting had been used in Massachusetts in the 1850s, but growers there abandoned it because they didn't consider it cost-effective.

One of the first growers to harvest on the flood was Andrew Searles, who cultivated native vines and developed the Searles berry variety that now makes up 65 percent of the Wisconsin crop. The McFarlin variety, developed in Carver, Massachusetts, accounts for another 20 percent.

Wisconsin briefly surpassed Massachusetts in annual production of berries in the mid-1970s. Now its growers produce 35 to 40 percent of the annual U.S. crop.

The Pacific Plantations

In 1874 Charles Dexter McFarlin built one of the finest bogs in Carver, but when a fall frost wiped out his first crop, he decided to return to the west coast where he and two brothers had traveled during the Gold Rush days. Charles's father had always gathered the wild native berries that grew in Carver's "New Meadows," and his brother, Thomas Huit McFarlin, had developed the McFarlin variety sometime before 1870.

Charles settled in what is now Coos County, Oregon, an area that reminded him of his hometown. He sent back to Carver for McFarlin vines, with which, by about 1885, he launched Oregon's cranberry industry.

In 1881 Anthony Chabot, a French Canadian, bought land near Long Beach in Pacific County, Washington. About two years later he began building the state's first bog, in what was still only sparsely inhabited territory. His nephew Robert carried the industry to the North Beach area in 1894.

Two Massachusetts men played roles in the early Washington industry. Anthony's brother-in-law from Massachusetts gave him the idea to grow cranberries and picked out the Long Beach site. Bion A. Landers of Cataumet provided the technical know-how needed to build a Cape Cod style bog.

Ed Benn built the first bog in the Grayland area in 1912 and soon sold tracts of land to Finnish settlers, who became a large part of Washington's cranberry industry.

In 1983 the Coastal Washington Research Extension Unit celebrated its sixtieth year of helping Washington cranberry growers. The station's first director was Daniel Crowley, who in the 1920s found that ice on a cranberry plant radiates heat inward, protecting the plant from frost damage. This proved to be a major discovery, which, with the perfection of the sprinkler system, greatly increased the growers' yields.

Joseph Stankiewicz, an Oregon grower, invented the Western dry-picker, and his ideas contributed to the development of the first water reel.

The McFarlin is the predominant variety grown in Washington and Oregon. The two states combined produce about 12 percent of the national cranberry crop.

Handpicking on Wisconsin grower Judge Gaynor's bog in the 1920s. (Ocean Spray Cranberries, Inc.)

Left: Hauling vine cuttings for a new bog in the Washington wilderness, early 1900s. (Ocean Spray Cranberries, Inc.)

Opposite page, top: Native Americans hand-picking a Wisconsin bog, circa 1890. (Ocean Spray Cranberries, Inc.)

Opposite page, bottom: Harvesting floating berries using the Wisconsin-style scoop, 1960s. Unlike the classic Massachusetts cranberry scoop, which was used by pickers on their knees, the Wisconsin scoop is used by pickers on their feet. According to a report of the Immigration Commission in 1913, this method "...gives better results in every way and is not nearly so fatiguing..." (Ocean Spray Cranberries, Inc.)

35

No individual has made a stronger mark on the cranberry industry than did A.D. Makepeace. A brilliant businessman and an innovative cultivator, he was one of the first to envision cranberry growing on a grand scale. A.D. saw the future in large bogs and centered that future in Plymouth County. As a founding member of the Cape Cod Cranberry Growers' Association, he was among those who saw that working together was the key to harnessing the future. A.D.'s time spanned the eighteenth and nineteenth centuries. He was a true bridge between old and new. (Ocean Spray Cranberries, Inc.)

A. D. Makepeace

Megan Tarini

Abel Denison Makepeace was born in Middleboro in 1832. His father, Alvin, was a cloth manufacturer, and his grandfather, Deacon Lysander Makepeace of Norton, had been a state representative. In 1854, the 22-year-old A.D. bought land in Hyannis and became a farmer. Two years later he married Josephine Crocker, who bore him five children, only three of whom, William, John C. and Charles, survived to adulthood.

A.D. grew potatoes and strawberries on his farm, but like so many Cape men in the 1850s, he caught cranberry fever and by the 1860s had built a small bog. At first unsuccessful, by 1871 his bog was producing over 2000 bushels of cranberries, leading him to comment in a letter to Joseph J. White of New Jersey that, "...the profits of cranberry culture are usually large...beyond comparison with any farm crop raised about here..."

With the financial backing of George F. Baker, a cranberry commission merchant who had "...unlimited faith in the business sagacity of Mr. Makepeace..." (Deyo, *The History of Barnstable County*) A.D. built the Newtown Bog in Barnstable in 1874 and a few years later started the Wankinco Bog in Carver. Wankinco, at 160 acres, was to become the largest individual bog in the state at that time. (Sullivan, *Cranberry King*)

By 1879 A.D. would be using his own capital to expand his enterprises in Barnstable and Plymouth Counties, but his relationship with Baker remained strong. In 1896 the two joined in an effort to bring electric trolleys to the Cape, but unlike most of A.D.'s schemes, this was one idea whose time was not yet ripe. (Sullivan)

In 1879 the Makepeace family moved to West Barnstable. Here A.D. could better oversee his operations, which now included a cooper shop and the West Barnstable General Store (which he had bought primarily so that his laborers could buy supplies on credit). His holdings continued to grow throughout the 1880s. The Wankinco Bog Company was organized in 1882, the Frogfoot Bog Company in 1885, the A.D. Makepeace Company in 1886, the Marstons Mills Company in 1888 and the Woodland Company in 1889. Two other companies, the Mashpee Manufacturing Company and the Carver Bog Company, had been organized in the 1860s.

Other growers owned shares in these companies and served as senior officers. A.D., as was his custom, preferred to be treasurer and general manager, which he believed gave him greater control over his investment. (Sullivan)

One of A.D.'s biggest successes was not in cranberries but in brickmaking. In 1887 he and several partners acquired the West Barnstable Brick Company. Under the management of B.F. Crocker as president, A.D. as treasurer and A.D.'s son William as secretary, the company's yearly output jumped from a few hundred thousand to two million within two years. William, who also ran the general store and the Cape bogs, continued to run the brick company until it was sold in 1926. Five years later it went bankrupt.

By late the 1880s, it became clear to some growers that a forum was needed for the exchange of ideas and information on cranberry cultivation and marketing. To answer this need, the Cape Cod Cranberry Growers' Association was formed in the summer of 1888, with A.D. as a founding member. The purpose of the organization was to "propose and discuss any measures deemed to be for the benefit and advancement of the business." Its first officers were John J. Russell, president; Emulous Small and A.D. Makepeace, vice presidents; and Isaiah T. Jones, Secretary.

The stature A.D. had gained among cranberry growers and merchants by this time was not unearned, for in spite of his diverse interests and holdings, he was always a cranberry man, committed to growing and shipping only the best berries. To that end he made improvements on the snap scoop and devised his own picking devices; he built his own mechanical separators and set high standards in his screenhouses; and he built his own barrels in his cooper shop. He also wrote a book on cultivation in 1884 (which has since been lost), and not long before his death he and his son John C. began experimenting on dehydrating berries.

A.D. expected and got full return for his efforts. Ac-

cording to Sullivan in *Cranberry King*, "Once he got what he wanted he became used to it and never wound up on the short end of a deal." Yet though he was a tough businessman, he was also a fair employer by the standards of the time, who understood the necessity of maintaining a contented workforce. This he did by extending credit at the general store and providing housing for his employees, and by such simple gestures as bringing homemade snacks to his bog workers and providing new shoes for workers whose shoes had been destroyed by fertilizer. (Sullivan)

[One day during the spring sanding season], several of the men, feeling they were getting paid too little for the amount of work they did, decided to carry only enough sand to correspond to their lowly pay. That morning A.D. Makepeace was in the bog and as the men wheeled by him with their load of sand, they all greeted him with a "Good morning, Mr. Makepeace," and he would respond with the same "Good morning" and the man's name. Every once in a while he noticed a man come by with a small load of sand, and after several had gone by, he said to the next man with a small load, "Good morning,

small load," and the small Portuguese worker replied calmly, "Good morning, small pay." Makepeace was surprised and, a week later, each man was given a pay raise. (Manuel Roderick, quoted in *Cranberry King*)

In 1911 A.D. was 79 years old and had been slowing down for some time. Age may have been the reason he merged all his cranberry companies that year with A.U. Chaney's New England Cranberry Sales Company. It was not a happy decision for him, but perhaps his instincts told him, correctly, that the future of cranberries was in cooperative marketing.

A.D. had invested in bogs in New Jersey in 1902 and around the same time had bought orange groves in Miami Beach, Florida. These purchases set a pattern for his later years. During the picking season he lived in Wareham to be near the bogs, and when the harvest was over he and his wife traveled to Florida for the winter. In the spring they stopped in New Jersey to check on the Makepeace holdings and spent the rest of the spring and summer in West Barnstable.

In April 1913, on his way north, A.D. fell ill in New Jersey with pneumonia. Although he came close to overcoming his illness, on June 23 he died at his home in West Barnstable.

Building a Cranberry Bog

Robert Demanche

The wild American cranberry, *Vaccinium macrocarpon*, is a trailing evergreen vine found as far north as Newfoundland, west to Minnesota and south to the higher elevations of North Carolina. Cultivated berries, larger than the native stock, are grown in Massachusetts, New Jersey, Wisconsin, Washington and Oregon, as well as British Columbia, Ontario, Quebec and Nova Scotia and to a small degree in Rhode Island and Connecticut.

The "runners" of the cranberry plant can be up to six feet long with uprights a few inches high, and its dark green, glossy leaves form a dense mat over the bog surface. The leaves turn a deep burgundy in the fall and winter when the plant is dormant. In the spring, the leaves start to "green up," and by late June the whitish-pink flowers make their appearance. Most of the fruit is borne by blossoms on the uprights. The berry reaches maturity 6 to 12 weeks after full bloom, depending on the variety.

The cranberry's roots are fine and fibrous. They develop from the runners and extend through the upper few inches of soil. The roots do not have root hairs, but instead absorb the required nutrients through closely associated fungi.

Some historians suggest that the slender stem and the downward-hanging blossom, because they resemble the head and neck of a crane, gave rise to the name "craneberry," later shortened to "cranberry."

In 1988 the Early Black and Late Howes varieties accounted for 97 percent of the Massachusetts crop. Early Blacks are popular for their ability to produce on many types of cranberry soil. Maturing in September, they are blackish-red when ripe, pear-shaped and quite uniform in size. Howes berries mature by about early Oc-

"Upright Cutting Planting." An engraving from Eastwood, The Cranberry and Its Culture, 1856.

tober, turn medium-red when ripe and color well in storage. They are glossy, oblong berries, slightly larger than Early Blacks.

Selecting a Bog Site

In the early 1800s, cultivators tried to improve the yield of the native vines they found growing in low-lying meadows and swamps and along the sides of ponds. They experimented variously by transplanting vines, adding layers of different types of soils, keeping the soil at different degrees of saturation and damming up streams to provide frost protection. Several growers, mostly in areas outside Plymouth and Barnstable counties, claimed to succeed at growing cranberries in upland rather than lowland areas, but this method found few followers as the industry grew in Massachusetts. Some early growers

felt that the best "cranberry yard" sloped toward the north; others preferred a southerly facing slope.

In 1856 Benjamin Eastwood of Dennis published the first manual on cultivation, *The Cranberry and its Culture*, which included a brief history of the industry and letters on growing techniques from growers throughout the state. Through Eastwood growers learned that cranberry vines grow best on a coarse, sandy soil above the rich peat or muck layers formed by decaying vegetation. They also learned that a grower did not plant cranberry seeds, but instead secured cuttings of healthy vine. He pressed the middle portion of the vines into the ground so that roots would form and the vines would cover the ground in a thick mat.

Eastwood instructed growers to add layers of sand to their yards every few years and to maintain a controllable supply of water to flood and drain the bogs for protection against cold winds, sudden frosts and insect infestation. He suggested dividing bogs into sections isolated by earthen dikes and allowing the water to flood the bog via flumes in the dikes. Bog ditches were to carry the water on and off the bog and the bog surface was to be a foot or more above the water table.

In choosing a bog site, nineteenth-century novices were guided by such advice as:

> ...if you plant on the upland it is difficult to raise your vines to bearing perfection, and it will entail much labor and expense upon him who undertakes it. Guard against the upland mania. (Eastwood, 1856)

> A huckleberry, maple, or cedar swamp [is preferable] to a fresh meadow, for the reason that it costs less to care of it after it is

planted. Less grass and fewer weeds will grow in a swamp after it is first cleared, than in a meadow. The swamp on which nothing but wood has grown, has the best bottom; it is enriched by the decayed leaves, etc., of years, and no nutriment has ever been taken from it by other vegetation. (James Webb, 1886)

Much of the land that eventually became cranberry bog, especially in Plymouth County, was originally swampland, by nature full of the essential peat or muck that provides nutrients to the vines. Drained mill ponds and sand-covered salt marshes and meadowlands also made good growing locations. An open location posed less of a problem with frost than one shut in by high uplands and woods.

Today there are strong environmental pressures against growing in wetlands, forcing some growers to reconsider upland bogs. But upland construction costs more. It requires the builder to excavate an area and make a level gravel bottom, and then apply a peat layer and an upper sand layer. The bottom line for the grower who wants to build a bog today is this: He or she must be willing to invest $20,000 per acre and wait four years before the first full crop is ready.

Left: An engraving of a cross-section of good cranberry land—swampland rich with muck and surrounded by coarse sand. "Swamplands upon which the White Cedar, or Juniper, the Maple, Swamp Huckleberry and Magnolia thrive, are [where] the deep deposits of muck are found, and, when properly prepared they make lasting and valuable meadows." Joseph J. White, Cranberry Culture, 1870.

Below: Cedar and pine have been cleared in building this cranberry bog on Cape Cod, ca. 1945. According to Farmer's Bulletin, No. 1400, published in 1924, when stumps are cut below ground level and covered with peat and sand, their roots help in drainage, thus improving the soil quality and saving the grower considerable expense. (Ocean Spray Cranberries, Inc.)

Opposite page: The cranberry blossom, showing its close resemblance to the graceful profile of a crane. (Ocean Spray Cranberries, Inc.)

Clearing the Land

Should a swamp be the chosen spot, the first step will be to cut a main ditch to the depth of two feet below the surface of the muck underlying the turf and roots, and of sufficient width to prevent the surface from becoming flooded with heavy rains. This drainage will enable the remainder of the work to be done more expeditiously, and more cheaply. (Joseph J. White, 1870)

Often a bog suffers from being too wet rather than too dry. Poor drainage promotes weed growth and some harmful insects and rot diseases. It also curtails root growth.

In clearing a bog, we first mow off all the bushes and low undergrowth. Next, we cut the principal roots of the large trees, and the wind will generally blow them over. The ground being soft, they will pull up a great mass of roots in their fall. This is the cheapest and best way of getting rid of trees.... The trunks are sawed up into logs for boxboards, or cut up into cord wood, or given away. The refuse tops and brush are piled in heaps and burned on the bog, as soon as they are dry enough. (James Webb, 1886)

Land clearing using turfing hoes and other hand implements remained much the same until just after World War II, when heavy machinery became common. Whereas it might have taken years to clear land for five acres of bog in the old days, today bulldozers and backhoes have shortened that process to two months.

After clearing the land of trees and stumps, turfing, or scalping, came next. This rid the land of the top layer of weeds and grasses that compete with the cranberry vine for nutrients, water and sunlight. The builder used an axe or cleaver for this job, as well as a turfing hoe.

A turfing hoe is made of plate-steel, about six inches wide and ten inches long. Before using the hoe, however, the dexterous workman cuts his turf in strips, twelve inches wide, with a cleaver or ax. The average cost of turfing is now twenty-five cents per square rod, or $40 per acre. The next step, after loosening the turf, is to remove it; and to accomplish this, the "floats" may be conveyed to the shore.

Grading

Below the top layer of turf lay the soft, nutrient-rich peat layer. The growers wanted a level peat base on which to add a layer of sand. By the 1880s, they used a grading hoe ("Every farmer knows what that is..." according to Webb) to achieve a smooth surface.

A bog should be graded and made as smooth and level as possible—in fact, as near a "water level" as it can be made—because if it is of a good, even grade, it will take less water to flood it. (James Webb, 1886)

Modern bog builder Irv Howes says he was the first on the Cape to use laser technology to grade cranberry

A stump winch, ca. 1925. The 3hp engine that powered the stump winch was geared so powerfully that 3 or 4 stumps could be pulled at one time. By hand three men might remove 14 to 20 stumps a day; with a stump winch, they could remove 125 stumps in an hour. (Ocean Spray Cranberries, Inc.)

Left: A clamshell digger, 1946. These were the only heavy machines used for bog building prior to World War II. (Ocean Spray Cranberries, Inc.)

A modern bulldozer equipped with a laser to ensure a perfectly level bog floor, Morse Swamp Bog, 1989. (John Robson photograph)

bogs in 1982 and he was one of the first to use a wide-track bulldozer on the soggy peat surfaces. The laser has since become standard equipment, because it can keep the surface within a quarter of an inch in level over a distance of a thousand feet.

> The laser itself puts out a beam, a reference point, over a 360-degree range. You pick up this beam in the sensor on the front of the bulldozer. The sensor is hooked up to the bulldozer blade. If you see a green light up from the operator's compartment, that means that blade is exactly where you want it. You just control the blade yourself, by hand. So the minute you move the blade, you either get a light that says you're too high or too low. (Irv Howes)

Heavy equipment allows modern-day bog builders to carve out perfectly square or rectangular bogs and to remove many of the large boulders or stumps that in former days would have been too difficult to pull out with the growers' shore-based winches. Many older bogs have a characteristically irregular shape that followed the outline of the natural low-lying ground. A neat square or rectangle was just too difficult to cut out of the higher surrounding area using hand labor.

Mining sand for a new bog, ca. 1915. This backbreaking job had to be done by hand because most bogs were built in areas too swampy to support draft animals. According to grower Clark Griffith, when hauling sand, "...the last thing you did before you left the pit was spit on your hands." (Ocean Spray Cranberries, Inc.)

Sanding

After grading the next step in bog construction is sanding. Dr. Franklin explained in 1915 why the top layer of sand is essential to the growth of the cranberry vine:

> It helps to keep down weeds and moss; it fastens down the runners and enables them to root better; it gives the roots a medium to grow in which is capable a far better drainage and aeration than is peat, and thus promotes their greater development; it takes in more heat during the day than peat, and radiates it at night so as to afford a considerable protection from frost; it is a considerable aid in controlling some insect pests.

Sand is always screened first. In the early years stones were hard on pickers' knees, and today rocky sand can harm dry picking machines. Joseph J. White described how to determine what type of sand was best for cranberry growing:

> Take a portion of the soil and compress it tightly in the hand; if it is suitable, it will fall apart upon being released; but if composed in part of loam, it will adhere together after the pressure is removed. This is a simple but reliable test, and one much used by practical growers.

Irv Howes explains the reasoning behind White's test: "If it sticks together, the water will stay in it — you won't get any drainage."

Sand is abundant around the swamps of southeastern Massachusetts. In the early days, ca. 1910 here, it was carted in wheelbarrows over movable planks. Care was taken to spread the sand to a uniform depth of 3 to 4 inches so as not to mix it into the peat layer and thus upset the soil structure. (Courtesy of Larry Cole)

By the 1940s, trucks and elevators were used to mine and haul sand. (Ocean Spray Cranberries, Inc.)

Opposite page, right: Screening sand to remove stones, ca. 1940. (Ocean Spray Cranberries, Inc.)

Planting in furrows, a job usually performed by women in the early days. According to White, "One woman...drops the vines in drills [furrows] while another follows her with a hoe." The women got 75 cents a day for this work, which cost the grower about $8 an acre. From Cranberry Culture, 1870.

Top right: A marker. Drawn across the bog surface, the marker created a checkerboard of guidelines for planting. This design, set on a 9-foot-long, 2- x -4-inch joist, had teeth 8 inches long and 18 inches apart. Illustration from James Webb's Cape Cod Cranberries, 1886.

Below: A four-pronged dibble. This was not a common model; perhaps it was an imaginative grower's failed experiment. (Ocean Spray Cranberries, Inc.)

Below right: "Dibbling in cranberry cuttings." From Corbett, Cranberry Culture, 1903.

Planting

According to Eastwood, the oldest method of planting or transplanting, used by Henry Hall and other pioneer growers, lasted only a few decades:

> It was customary to cut out a square or oblong sod on which the vine was growing, and then to prepare the yard to receive it just as it was taken up. But experience has taught cultivators that this is not the best mode. In removing the sod, rank weeds and foul grasses are brought with it, [which] retard the matting process of the vine, and the yard becomes one of weeds and wild grasses rather than of healthy cranberry plants.

In 1886 James Webb described the method most growers came to favor:

> New cranberry meadows are almost always established by planting cuttings, 10 to 15 inches long. The common practice is to secure the cuttings from vigorous plants by mowing a portion of the meadow with a mowing scythe...and separate them into wisps containing from 8 to 15 stems. The wisps are placed at [intervals] 18 by 18 or 9 by 18 inches. The cuttings are then forced into the sand with a broad, thin wedge-shaped dibble. The blade of the dibble is placed midway of the wisp of cuttings, so that the pressure exerted upon the cuttings doubles them upon themselves and at the same time presses them firmly in the soil.

Many years earlier, Eastwood had confirmed this method "good and safe."

Today growers plant by spreading the cuttings by hand onto the bog surface and pushing them into the sand with a mechanical vine-setter. This machine drives a row of several steel disks, 6 to 8 inches apart along an axle, that press the cuttings into the sand as the machine plows back and forth across the bog surface. Immediately after planting, water is applied for a day or two to wet the vines and allow the sand to settle in around them. The newly planted vines bear some fruit as they "run" across the bog during the first year of growth, but it isn't until about the fourth year that the vines completely cover the surface and a full crop can be harvested.

Hand-planting under the watchful eye of the overseer on the Allen-Bradley Cranberry Company's Charge Pond Bog in Plymouth, ca. 1900. Carver grower Wilho Harju recalls that in his early days he and a crew of 14 people could plant up to 1-1/2 acres a day with dibbles. (Ocean Spray Cranberries, Inc.)

Left center: A hand dibble. This is the more common type. The wedge-shaped blade pushed the mid-section of the vine cutting firmly into the soil so that it touched the peat layer. (Ocean Spray Cranberries, Inc.)

Bottom left: A properly dibbled vine cutting, standing an inch or two above the sandy surface. From Corbett, Cranberry Culture, 1903.

Using a marker. A cutting is planted in each corner of each square created by the marker. Before 1920 all cuttings were set 18 inches apart. Since then, growers have been setting them closer, especially on the richer swamplands. The closer set, the better they anchored themselves against the pull of picking scoops. Close-set vines also cover the ground sooner and require less weeding.

45

Carver grower Clayton McFarlin (1873 - 1962). Clayt started out as a surveyor. He applied the precise approach and skills demanded by that trade to raising cranberries on 26 acres of bog that were beautifully kept and consistently high-producing. Before World War II, when most growers were satisfied with 50 barrels an acre, Clayt raised over 425 barrels an acre on his Paul Bog one season. He would complain to his box supplier that he had so many berries that he didn't know how he could get his entire crop under cover.

Clayt's bogs may have been the picture of high-yielding health, but his screenhouse at Huckleberry Corner was something less: an abandoned barn with several carriage sheds attached. He never got around to building a new screenhouse, but his bogs continued to produce the biggest and best crops in the cranberry industry. (Middleborough Public Library; caption notes by Larry Cole)

Gary Florindo, driving a bog buggy rigged with a vine-setter in the rear on Joseph Barboza's Morse Swamp Bog, 1989. (Joseph D. Thomas photograph)

Paul Fernandes, Howard Pierce and Todd Barboza, spreading cuttings in advance of the vine-setter on Joseph Barboza's Morse Swamp Bog. (Joseph D. Thomas photograph)

A knife-rake for pruning, made by H.R. Bailey, ca. 1940. The razor-sharp teeth are detachable for sharpening. (Ocean Spray Cranberries, Inc.)

Pruning at Ellis D. Atwood bogs, 1953. Pruning was necessary after harvesting to straighten out the loose runners brought to the surface during scooping. To make harvesting easier, the extra runners were cut by experienced pruners who took extreme care not to damage the vines. Today dry-harvesting machines prune as they pick, and wet-harvested bogs do not require pruning. (Ocean Spray Cranberries, Inc.)

Bottom right: Cutting weeds, ca. 1960. This grower is using a gasoline-powered cutter for clipping weeds. The weed tops are cut into small pieces that fall down among the vines and do not have to be removed from the bog. (Ocean Spray Cranberries, Inc.)

Pruning and Ditching

For the first two or three years after planting, Joseph White advised regular weeding of grass and other unwanted vegetation in order to give the vines "undisputed possession" of the bog. Eventually, however, vines grow so thick and tall that a periodic pruning is necessary to take off unproductive overgrowth.

For decades growers pruned the overgrown area with knife-rakes, but today dry-harvesting machines are rigged with a cutter blade that prunes automatically as the picker moves over the bog. Water-harvesters may prune their bogs every five or six years with a pruning machine.

Ditches must be cleaned manually or with mechanical ditch diggers. Today, the dredged-up weeds and mud are wheeled off or set out on mats to be "mud-lifted" away by helicopter. Grower Eino Harju says that traditional shoveling is still the best way to clean a ditch, although it is almost impossible to find workers to do it.

Clearing brush on Atwood Bogs, South Carver, 1989. Jose Dias works with David Eldredge Jr., clearing and burning brush on some of the 200 acres of Atwood bogs managed by David's father. (John Robson photograph)

Bog ditches must be weeded and cleaned regularly to maintain proper water flow and drainage. The old way of removing weeds and sediments was with potato-diggers and shovels; the cleanings were piled beside the ditch and left to drain for a few days, then loaded on a wheelbarrow and carried ashore. Usually done in the spring or fall, this work is often cold, wet and dirty. (Ocean Spray Cranberries, Inc.)

Today helicopters lift the mud mats as they are piled along the ditch and drop them on compost heaps on the shore. (John Robson photograph)

Flooding

Successful cranberry cultivation depends on an abundant supply of water. A bog must be flooded or sprinkled periodically for irrigation and for protection against frost, winterkill and insects. To regulate the amount of water flowing onto or off the bog, bog builders have historically constructed ditches, dikes and flumes.

The first step is to build a marginal ditch, 3 feet wide and 2 feet deep, around the inside edge of the bog. According to Franklin, this ditch "prevents upland growths from working onto the bog, keeps many crawling insects off, and is some protection from forest fires." Cross ditches, which divide the bog into sections, are necessary if the bog is large or the drainage is poor.

Dams, or dikes, surround the bog to hold in the water and separate bog areas of different elevations to promote level flooding. Dikes are built of sand, narrower at the top than at the bottom, with turfed sides. Growers have always used the top of dikes as roadways.

In 1915, Dr. Franklin cautioned growers:

Heavy teams should not be allowed to drive over a new dike for several months after it is built, for the dike will be injured for holding water if it is used as a road way before it has become settled together.

Model A's, pickup trucks and tractor trailers have all traveled over bog dikes. Irv Howes talks about the hazards of using modern vehicles on old dikes:

> The roads have to be built a little stronger, a little wider to pick up the cranberries in the fall harvest. They collapse, and some are so narrow, you can't open your door without falling out. I think everybody in the industry has lost a truck or something off a dike.

The passage of flood water from a reservoir to the bog or from one bog section to another can be controlled by flumes—openings built into a bog dike. A flume consists of a gate made of several 6-inch wide boards, called flume boards, stacked horizontally on end and inserted through tracks, or keyways, running vertically along the flume's inside wall. The flume boards are removed according to the water level of the adjacent water supply.

In 1886, James Webb described the popular system of the time:

> The gate, constructed of matched inch boards, slides up and down between strips or grooves of wood…by means of a chain and roller. This is one way of constructing the gate and it is the common one on the Cape. (James Webb)

In early years, when gravity provided the only force for flowing water, growers built their bogs at elevations lower than the water source. When power-driven pumps

Nineteenth-century flume gates, used to control the waters of the East Head Reservoir for Big Bog on the Wankinco watershed. The old wooden planks used to form the portable flume gates are employed just as they were 100 years ago. In many of the modern dikes, however, flumes are one piece of standard aluminum culvert pipe instead of the redwood, stone or concrete structures used in earlier days. (John Robson photograph)

Below: A pump house at the Big ADM Reservoir along the Wankinco, near the Wareham-Carver line. (John Robson photograph)

became available, growers could locate their bogs near reservoirs at lower or equal levels. Dr. Franklin wrote in 1923,

> Formerly, single-cylinder gasoline engines alone were used in bog pumping plants, but lately multiple cylinder engines and electric motors have found much favor. Propeller pumps are best for lifts less than 4 feet, but reversed turbine pumps are better for lifts of 5 to 12 feet.

Early examples of Yankee ingenuity in power-driven pumps were steam plants and windmills, but perhaps the most ingenious idea was a horse and capstan connected to a series of buckets that picked up water from a pond and dumped it into a trough leading to the bog.

Low-gallonage sprinkler systems built into the bog surface were introduced in the early 1960s. They have proven very effective for frost protection because they efficiently disperse enough water to encapsulate the plant in a protective casing of ice. They are also used for irrigation and pesticide application.

Sprinkler systems have not yet replaced the need for flooding, however. Following the harvest, if the grower has adequate water he will fill his bog with a short "cleanup flood" to water vines damaged by harvesting machines and to float dead leaves and bruised berries for easy removal. He also floods his bog again in December, and keeps it flooded until March, to protect against "winterkilling," a condition that exists when sub-freezing temperatures, winds of at least two days duration and frozen soil in the root zone combine to prevent the vine from replacing the moisture it loses through its leaves. Serious winterkill occurs about once every three years.

A "late-water" flood in the spring produces larger, higher-quality berries and reduces the numbers of some pests. But it uses valuable water, produces fewer berries and does little to slow the development of many weeds and insects. In the summer, extreme heat can cause desiccation. A quick flooding at night or, better yet, turning on the sprinklers can counter the heat's ill effect.

The developing cranberry's tolerance to frost conditions changes throughout the growing season. The bud can stand low temperatures, down to –15°F, in winter as long as they are not accompanied by moisture-robbing winds. In spring it will tolerate temperatures down to about 18°F. As the bud develops into a berry, it becomes more tender and less able to withstand temperatures slightly below freezing; in summer 28°F is the minimum. As temperatures drop in autumn, the ripening berries are able to endure a moderately colder temperature, but the grower must still watch out for the autumn frosts that can happen whenever temperatures fall below 30°F.

Before the introduction of the low-gallonage sprinkler system, the only way growers could protect their crops from frost was to flood the bog overnight, making sure that all the vines were completely submerged. They had to know how long it would take to fill a bog and when to start flooding to beat the coming frost; if they miscalculated, flooding after the critical temperature was reached was practically useless. The sprinkler's rapid response time and efficient use of water has made frost much less of a problem and has thus promoted higher yields.

52

An old flume gate, used to control the waters of the East Head Reservoir for the Wankinco bogs of the A.D. Makepeace Company, Miles Standish State Forest, South Carver, 1989. (John Robson photograph)

Top and bottom left: Sprinklers at work near Morse Swamp, 1989. (Joseph D. Thomas photographs)

Opposite page, top: Ronald Amaral, lifting flume planks with a plank hook. The plank hook is one of the least modified tools in the history of cranberry culture. It is used to hook a steel staple embedded in the plank. The original plank hooks had a maritime, double-ended gaff shape, which may have had something to do with the fact that most early growers were seafaring men. Today's plank hook's look like a vulture's beak. (John Robson photograph; Carolyn Gilmore caption)

Opposite page, bottom: A spillway releases high waters from an abundant reservoir, ca. 1958. Many growers rely on excess water spilling over from local reservoirs and rivers for their water supply. For this reason, the heavy snows and steady rains of winter and spring are a blessing for most growers. (Ocean Spray Cranberries, Inc.)

Resanding

Though costly and laborious, spreading 1/2 to 3/4 inch of sand on a mature bog every few years is considered an essential part of good bog management. Resanding provides new area for growth and allows excess water to drain away, leaving air spaces that deliver oxygen to the roots. Also, a sand covering accelerates the breakdown of the "trash layer" of fallen leaves into nutrients taken up by the roots. Covering the trash layer helps to make the "micro-environment" near the vines unsuitable for large-scale infestations of some insects.

Because sand encourages growth, by the second year after sanding, the vines shade the bog surface enough to prevent many weed seeds from germinating. And though not proven, the sand is thought to act as a reflector, training a second dose of sunlight onto the plant.

Finding sand has always presented much less of a problem for growers than moving it. Early growers sometimes used oxen to cart sand or, more often, had their workers shovel the sand into wheelbarrows and wheel it over planks to the bog. On some larger bogs, workers laid down temporary tracks so an engine could pull small cars loaded with sand over the greater distances.

Today, most growers use homemade sanding trucks and drive back and forth over the bog releasing sand from a hopper in rear end. Some growers have built boats and sand directly in the flood water. Helicopters have also been used for sanding, but this has not become common practice.

If the winter flood freezes over, growers can dump and spread sand directly onto the ice. When the ice melts, the sand distributes itself evenly over the vines. Ice sanding eliminates the damage to the vines caused by workers and machines.

Left: Howard Hiller of Rochester, inspecting his sanding gear. The sanding cars are pulled by H.R. Bailey's locomotive. Bailey designed these engines after those used for hauling coal up from a mine. They traveled over the bog on moveable tracks. (National Archives)

Below: Helicopter sanding on A.D. Makepeace Company's 13 Acre Bog. On the larger bogs, sanding by air is becoming more popular. (John Robson photograph)

Opposite page, top left and right: Trucking sand over planks with Bailey diamond sandbarrows. Sanding started at the shore and worked toward the main ditch. The sanding crew's wages often included a few cents for each 15-foot plank they had to travel before they dumped their loads. According to grower Larry Cole, it wasn't uncommon for the boss to have two long-legged men leading a column of sand wheelers and one aggressive, ill-tempered wheeler pushing the less ambitious men from behind.

Forged by H.R. Bailey at his blacksmith shop in South Carver, the diamond sandbarrows were specially designed so that the bulk of the load was over the wheel for balance. On the bog, the sanding gang would turn the sandbarrows so that the wheel was on the plank and the legs were on the bog. A well-placed flick of the shovel ensured good spreading. In the thirties the pneumatic steel tire made the barrow easier to push and eliminated the need for planks because it didn't injure the vines. (Photographs courtesy of Ocean Spray Cranberries, Inc., caption notes by Larry Cole and Carolyn Gilmore)

Opposite page, bottom: Ice sanding by Model A truck on one of Alex Johnson's bogs, ca. 1926. The man spreading the sand is Kulta Lahti. The sanding truck has a self-propelled tip cart; its drive mechanism can be seen on the rear left wheel. Carl Urann was the designer.

Maurice Fuller of Rochester says he was one of "six or eight around town" who contracted ice sanding services in 1940. With his dump-body rigged Model A, he went from bog to bog, earning about $5 a day after expenses. "During the week, when I was in school, I'd pay my uncle to drive and still make $2." (Ocean Spray Cranberries, Inc.)

Chemicals

For well over a century after Henry Hall set out his cranberry yards, cultural methods such as resanding and flooding were the growers' main weapons against pests. That changed in the early 1900s when pesticides and herbicides became available, which have helped the grower produce more berries by cutting down insect populations and destroying choking weeds. As these chemicals has become more readily available, the cost of pest and weed control has come down.

Before effective chemicals, handpulling was the only way to get rid of weeds. It was grueling for the workers and costly for the grower, but it was also effective. Handpulled bogs were freer of weeds than are bogs treated with herbicides today.

Some early growers spread ashes on their bogs to fight insects; others experimented with steeped tobacco, lime and Bordeaux mixture. Some growers applied what few pesticides were available with knapsack sprayers. These contraptions, used only in small areas, required one hand to work a pump, one hand to hold the spray hose and applicator, and a sturdy back to carry the pesticide "knapsack." In some areas growers used a horse-powered pump. They loaded barrels of pesticide onto low-bodied, horse-drawn wagons and two men sprayed using pressure supplied by the horse's motion.

Shore-based spraying made pest control easier on large areas. Manpower or small engines pumped the liquid from mixing barrels through long hoses carried by several men across the bog. Later, motorized power dusters and sprayers were used.

Growers added more powerful chemicals to their arsenal by the late 1930s, among them copper sulfate, pyrethrum, lead arsenate, sulfuric acid and sodium cyanide. After World War II, petroleum-based chemicals such as kerosene and Stoddard solvent were found to be effective herbicides that did not harm the cranberry plant, and DDT came into use for fighting insects. Obviously, while these chemicals helped the grower produce more, they had serious health and environmental drawbacks.

Movable platform for concocting Bordeaux mixture. Around 1900, Bordeaux mixture was considered the best fungicide. According to C. L. Shear, author of the 1905 Farmers' Bulletin, No. 221, the recipe for Bordeaux mixture, simplified here, was equal parts of slaked lime and copper sulphate, each mixed with water and then stirred together with a wooden paddle.

Right: Sodium cyanide mixture being applied with an electric, two-gear pump, ca. 1920. (Massachusetts Cranberry Experiment Station)

Chemical application by motorcycle, 1930s. Hayden, Bailey and others designed motorcycle sprayers. Hayden claimed that its model was "the best machine for the most efficient kill" in an advertisement in Cranberries magazine. (Massachusetts Cranberry Experiment Station)

Below: Spraying with a multi-nozzled rig, ca. 1930. (Middleborough Public Library)

In the 1930s, some growers with large, open bogs began dusting and spraying from airplanes. Experiments on aerial pesticide application intensified after World War II, and in 1947 the National Cranberry Association, prompted by Marcus Urann, purchased the first helicopter built specifically to spray and dust commercial crops.

Helicopters provided several advantages over straight-wing aircraft. They could cover irregularly shaped bogs well and maneuver in places airplanes couldn't reach, and they could land right next to the bogs for reloading. Moreover, the motion of the rotors helped force the dust to the floor of the bog for better coverage of the vines.

Growers eventually began applying fungicides and fertilizers by air as well, and by the mid-1950s, more than half of all cranberry acreage was treated this way. Truck and power sprayers and ground dusters accounted for one-third of the applications; hand-operated equipment accounted for less than 10 percent.

Today, two-thirds of all cranberry acreage is chemically treated using low-gallonage sprinkler systems.

Airplane dusting, 1940s. Airplanes proved to be too noisy and too hard to maneuver for today's bog-dusting requirements. (Massachusetts Cranberry Experiment Station)

The first public demonstration of the "Ocean Sprayer" or "NCA Machine" on June 5, 1947 at the 100-acre United Cape Cod Cranberry Company Pembroke Bog. (Massachusetts Cranberry Experiment Station)

Opposite page, top: A hand-pulled duster. (Massachusetts Cranberry Experiment Station)

Opposite page, bottom: Barry Zonfrelli applying herbicide Evital on one of the Federal Furnace Cranberry Company bogs. (Joseph D. Thomas photograph)

IPM

From about 1930, growers often sprayed for pests on the basis of their insect control and disease control charts and their own observations of bog conditions. Integrated Pest Management (IPM), a program developed in the early 1980s, has made the timing of pesticide application much more precise. Growers under IPM spray not because they see a few insects or because a chart says the time is right, but because, through regular monitoring of the bog, they have found a critical number of a certain species. More and more growers are joining an IPM program because it keeps chemical use down to the minimum necessary for effective pest control.

Growers are very careful about spraying while the cranberry vine is in bloom, since the chemicals might hurt the honeybees that are essential to pollination. Because the flower hangs downward, the heavy pollen grains it produces are not carried off by the wind but either fall to the ground or cling to the body hairs of a visiting bee. As the bee moves from flower to flower collecting nectar, some of the pollen it carries rubs off on receptive blossoms, aiding fertilization.

Growers may safely apply fast-acting insecticides through their sprinklers at night, when the bees are resting. In the morning, they turn on the sprinklers to wash away any residue before the bees start flying again.

About 20 percent of the bees used for pollination are local. Apiaries import the rest every spring.

Fertilizer

Growers apply most of their fertilizers in the spring and summer, when the blossoming cranberry requires energy for growth. The cranberry is unusual in that it will thrive on relatively small amounts of fertilizer; indeed, early in the twentieth century, Dr. Henry J. Franklin, director of the Cranberry Experiment Station, determined that most bogs did not need any at all. For some time, however, regular high yields have depleted soil nutrients, which can only be replenished by fertilizer application.

Maintaining a cranberry bog is a daily, year-round job. Machinery, pumps and irrigation systems require overhauling, roadways must be improved, flumes and dikes need regular repair, and the shore must be mowed to discourage weeds and insects. It takes hard work to keep a bog healthy, but a good harvest depends on it.

Part Two

Six Quarts to the Measure

Overview

Christy Lowrance

Early cranberrying was often a family operation. Only when the grower and his family could no longer manage the bog themselves did they hire first other family relations and then neighbors, community folk and eventually migrant workers. Men cleared the land and sanded it; women planted and weeded; and children helped with harvesting. Schools routinely did not open until October, and as late as 1927 permits were issued to enable children to miss school to pick cranberries.

In the early days down the Cape, you picked from the time you were six or seven years old. You might get a measure a day and get eight cents. Kids ten to fourteen would earn enough to buy their clothes. By the time you were fifteen, you were a man. That cranberry money was very essential. (Dud Eldredge)

Writing in 1935 in *Cape Cod Yesterdays*, Joseph Lincoln commented

Men and boys picked them—oh yes—but so did girls and women, married and single, and mothers and grandmothers. As for us youngsters, we all picked. No matter whether your father was as rich as Capt. Solon Crocker, who was chairman of the Board of Selectmen and who wore a beaver hat weekdays as well as Sundays and a fur neckpiece and gloves in winter, or as poor as Seth Cash, who kept his trousers up to the safety point

Opposite page: Pickers at work on an Ocean County, New Jersey, cranberry bog. Sketch by Granville Perkins, Harper's Weekly, November 10, 1877. (Courtesy of George Decas)

Handpickers at Eldredge Bog, ca. 1880. James Webb wrote in 1886: "When the season for cranberry picking arrives, it is no unusual sight to see nearly the entire population of the village, starting out in the morning on their way to the bogs....It is a sight which must be seen to be appreciated." Everybody brought his or her container for berries, lunch and a bottle of tea. (Courtesy of Doug Beaton)

A handpicker dressed for a day's work. James Webb found pickers' costumes to be "of startling originality and of the most unique description, the object being, not to see who can dress and look the best, but who can be the best protected in, and provided for the labor before them." This woman is wearing finger stalls to protect her hands from the vines. The ticket in her hat was for keeping track of the amount of berries she picked. From Country Calendar, November 1906.

Family wagon going cranberrying, 1911. Members of the Beaton and Gault family gather up for a ride over to 3 B's Bog in West Wareham. In the driver's seat is Thomas Gault and riding shotgun is Elliot Beaton. Behind them are Clara Gault Beaton holding Gilbert Beaton, a Guildaboni, Mrs. Gault, Alice Swanson Crane, Eleanor Beaton Morey, Missey Ellis and another Guidaboni. "We'd all pile into the wagon, lots of kids," recalls Eleanor Morey, "We'd take lunch and plenty of cold tea. I remember Elliot had fallen into the ditch when he was a boy and mother always told him that's how he got baptized." (Courtesy of the Beaton family)

with rope instead of suspenders—when picking time came, you were on hand.

Many first-person accounts of cranberrying reflect a nostalgic relish of the out-of-doors and the camaraderie experienced by the cranberry workers. Catherine Sears wrote about the end of a harvest day in 1920:

> The pickers, dog tired and feeling as if they were 100 years old, yet pluck up fresh spirit as they rumble along the homeward way in the old blue cart sing rollicking and jovial songs; for is there not before each one the dazzling vision of untold and well nigh inexhaustible riches which will be there at the end of the season's work.

Proper clothing was a critical part of picking, selected not for attractiveness but for protection from rough, wiry vines, damp ground and weather. Women wore sunbonnets and long denim skirts with oilcloth sewn across the front, while men wore heavy overalls with extra padding sewn into the knees. Cotton "finger stalls" protected fingers, and stockings over them protected hands and forearms. Shoemaker's wax protected the hands of those who didn't wear gloves.

Horse-drawn wagons and carts transported the workers to the bogs, which were usually located outside of town in seaside or wooded areas. These vehicles carried up to two dozen pickers loaded down with lunch buckets, thermoses and tin measures.

The days were long and the work was hard for these early cranberry pickers. On their knees, they moved up and down marked-off rows scooping up the berries from the vines with their fingers. Overseers spurred them on, always ready with a sharp reprimand for slackers. The call to "Knock off! Knock off!" signaled lunch, followed by a long afternoon back on the bog.

In 1913 one essayist described cranberry picking:

> The experience of the harvesters of the earlier days would now be regarded as hardship that would call for an investigating committee. The marshes were always damp, and in wet seasons, they were breeding places of rheuma-

tism and kindred complaints. The older laborers wisely refrained from a contact with the water, but the boys took no such precautions. Long lines of shivering barefoot boys lined out on the marsh, awaiting the signal for attack, and when the word was given they would drop into the icy water with shouts of laughter and boyish pranks, and the knees were numb with cold before the sun was high enough to impart its heat.

Joseph Lincoln probably spoke for thousands of cranberry pickers through the years when he wryly recalled,

Crawling back and forth for eight hours a day, rainy days and Sundays excepted, for a month or so, and then at the end of the season, being presented with four or five dollars, is not an easy road to riches.

Eventually, handpicking was replaced by scoops with wooden teeth and innovations such as hollow teeth "as big around as your thumb." Warner Eldredge remembers seeing a long-handled scoop with wooden teeth in a neighbor's barn when he was about ten: "I asked Nobel Swift how they worked. He said, 'Pretty good, if you was a man and used to shoveling.'"

By the 1890s, cranberry growing had become a major industry. More and more acreage was coming under cultivation and crops were getting larger, making it difficult for the grower to rely solely on family and neighbors at picking time. The days of the community harvest were coming to an end, to be replaced by a new era of immigrant labor. By the early twentieth century, two ethnic groups, the Finns and the Cape Verdeans, would make their mark on the cranberry industry of southeastern Massachusetts.

Family gathering, ca. 1890. (Ocean Spray Cranberries, Inc.)

Community Harvest

Joseph D. Thomas

Harvesting

...It is seldom that the best or quickest pickers gather more than three bushels during one day. To do this is extraordinary work. There must be a superintendent or overseer with them, or they will be apt to slight them. The interest of the cultivator is to have his vines picked clean. (Eastwood, 1856)

In former years, a grower would have the same pickers year after year, and they could be depended upon. But of later years, the pickers have been more independent, and, following in the footsteps of the great body of Knights of Labor—they strike. (O. M. Holmes, 1883)

Handpicking, with palms up and fingers spread and slightly bent, was the harvesting method least damaging to the vines and berries. But it was tough on the hands, knees and back, and it took an army of people to pick the bog clean. Nevertheless, the surplus of inexpensive labor prolonged handpicking; as long as people were paid by the measure and not by the hour, the number of pickers on the bog did not affect the cost of harvesting.

By 1880 the situation had changed. The pickers began to prospect the bogs to see which ones had the most berries. They would not work on a bog with a thin, poor crop, because they stood to earn much less than they could on a high-yielding bog. The increase in the number of bogs and a shortage of labor in the growing regions meant that pickers also began demanding higher prices for their output.

As the cape bogs are liable to an early frost, and the grower is anxious to get the crop under cover as soon as ripe, the pickers take advantage of the situation, and exact the best price they can get. (Holmes, 1883)

The growers' answer to the pickers' mounting independence was to tap the steadily growing immigrant la-

bor pool from the cities, and to develop a harvesting machine that would reduce the workforce and speed up production. In the 1940s, handpicking finally came to an end, overtaken by the speed of the cranberry scoop and the loss of pickers to the newly revived industrial and construction sectors of the economy.

Handpicking was well suited to the highly organized production line that cranberry harvesting had become in the late nineteenth century. However, before 1830 wild cranberries were considered public property, rather than a commodity in many towns, and adventurous growers harvested using primitive rakes, scoops and combs. In 1820 Barnstable sought to ban rakes on Sandy Neck, most likely because of the damage they caused.

Early cultivators used rakes sparingly, if at all, and then only to prune mature vines. To use them on young or matted vines "...would be folly," wrote Eastwood, be-

Competition between pickers could be fierce at times: "Nobody liked to pick beside Lizzie, for she was a 'scalper.' That means that she would get ahead of the one beside her and scoop all the top berries off, leaving the bottom berries for her partner. These bottom berries are the hardest to pick." From "The Cape Cod Cranberry Picker," by Geneva Eldredge, Cape Cod magazine, July 1915. (Ocean Spray Cranberries, Inc.)

"All the women and girls wore sunbonnets that concealed their faces from the sun's gleaming rays. Big blue denim aprons with a patch of oilcloth across the front enveloped them from the waistline to their toes." (Cape Cod magazine, July 1915; photograph from Ocean Spray Cranberries, Inc.)

Opposite page, top: Pickers on a Plympton bog, ca. 1890. (Courtesy of Karen Barnes)

cause the teeth of the rake would "do them serious injury." Aside from their damaging effects, rakes were also wasteful; many picked berries fell through the teeth of the rake and were trampled underfoot.

Henry David Thoreau, who did a little cranberrying himself, gave one of the earliest descriptions of water-harvesting in Massachusetts:

> I find my best way of getting cranberries is to go forth in time of flood, just before the water begins to fall and after strong winds, and choosing the thickest places, let one, with an instrument like a large coarse dung fork, hold down the floating grass and other coarser part of the wreck mixed with [it], while another with a common iron garden rake, rakes them into the boat, there being just enough chaff left to enable you to get them into the boat, yet with little water. When I got them home, I filled a half-bushel basket a quarter full and set it in a tub of water, and stirring the cranberries, the coarser part of the chaff was held beneath by the berries rising to the top. Then, raising the

Taking "floaters," the closet thing to wet-harvesting in Massachusetts before the 1960s. Floaters were unharvested berries salvaged when the bog was flooded after picking. Carver grower Eino Harju recalls how, as a boy in the twenties and later during the Depression, he hired himself out to growers at the end of picking season to take the floaters from their bogs. As was common back then, he got half the value of the berries salvaged and the growers kept the other half. Since floaters were sold only for processing, their market value was very low. (Middleborough Public Library)

A group of workers on a rugged Plymouth bog, equipped with an assortment of pails and pans for collecting the berries. (Courtesy of Larry Cole)

basket, draining it, and upsetting it into a bread-trough, the main part of the chaff fell uppermost and was cast aside. Then, draining off the water, I jarred the cranberries alternately to this end and then to that of the trough, each time removing the fine chaff—cranberry leaves and bits of grass—which adhered to the bottom, on the principle of gold-washing, except that the gold was that thrown away, and finally I spread and dried and winnowed them. (Thoreau, *Journals*)

In spite of Thoreau's endorsement, water-harvesting had little success in Massachusetts. In 1883 O.M.

Holmes, a Cotuit grower, reported that,

> This process requires the drying on frames, which is tedious, and consumes time and labor which more than offsets the cost of hand-picking. Berries that are gathered in this way are not as salable, and do not keep as long as those that are hand-picked. There were none gathered in this manner, in our section [Cape Cod], in 1882.

Picking in Staked Rows

...let the pickers all start in evenly, with instructions to keep in a straight line, which

A long-handled rake used in the mid-nineteenth-century for dry-harvesting berries for home or local consumption. (Ocean Spray Cranberries, Inc.)

Handpickers at a Rocky Meadow bog in Middleboro in the 1880s. (Courtesy of Francis LeBaron)

Handpickers in staked rows, ca. 1895. (Middleborough Public Library)

they can nearly do, for a short distance, by the slowest pickers taking the narrowest strips, and vice versa. (White, 1870)

Picking in rows was the best way to harvest the greatest number of berries. Growers "staked off" the bog, as Webb described:

...strings are stretched across the sections in parallel lines, some six feet apart, and made fast to pegs set in the ground at each end. Between these lines, the pickers are set at work, from one to three in a row, whichever they prefer; until his section is thoroughly and completely picked. It is best to keep strings ahead over enough ground to accom-

modate a gang of one hundred pickers, so that no confusion or delay may occur in setting them to work. (Webb, 1886)

The pickers were confined to their own row until all the berries were harvested. Friends or neighbors moved alongside each other and socialized while they worked.

Some care is necessary, at first, to properly discipline the pickers, and cause them to pick clean as they go. This may be done by calling them back in a pleasant, but decided manner, to gather any berries that may have been found after them. They will soon take the hint, and perform their work carefully. (White, 1870)

Workers picking in staked rows at a busy Harwich bog in the 1880s. Crowded bogs like this one were becoming more and more common as the industry grew in the late 1800s. No longer was the community sufficient as a source of labor; increasingly immigrants from Boston, Worcester, New Bedford and Fall River worked alongside the locals. According to one observer in 1936, "These nomads pitched their tents or erected rough cabins in the vicinity and went from bog to bog, as long as there were cranberries ripe for the harvest and not yet killed by the frosts." (Courtesy of A.D. Makepeace Company)

The Tally System and the Measure

As larger harvesting crews worked bigger bogs, keeping track of an individual's production became a tedious chore for the overseer. A standard picking measure was necessary that would simply and accurately determine the amount of berries each picker produced.

Sometime in the 1870s, the six-quart pail was introduced in Massachusetts. Before this pickers brought their own containers—usually kitchen containers such as colanders, baskets or tins. As they filled their containers, they collected the berries in burlap bags and had them counted up at noon or day's end by the boss, who recorded the total in a credit book. There were several problems with this somewhat casual system, as Joseph White noted:

> Much care is requisite, while picking, to secure the berries without bruising them. If they are poured into bags, and used for seats by the pickers, or thrown over their shoul-

GOOD FOR PICKING
6 QUARTS
OF CRANBERRIES FOR
F. D. Underwood.
CAHOON.

A tally ticket from the early 1900s. (Ocean Spray Cranberries, Inc.)

Top: Pickers turning in berries to the tallykeeper. On this Cape Cod bog, the grower (bearded man) appears to be overseeing the operation. (From Country Calendar, November 1911)

A tallykeeper recording a measure, ca. 1900. In Cape Cod Cranberries, James Webb wrote, "A measure can be filled by a smart picker in 15 minutes. A gang of 80 pickers (no unusual number) can in exceptionally good picking, pick 5 barrels, Massachusetts standard measure, in 15 minutes, or 20 barrels an hour, at which rate there must be an average of a credit to be given the pickers by the book keeper every 10 seconds. From this it will be seen that it takes a person of good ability, one who is quick and expert, to attend to the duties of accountant when the cranberry picking is lively." (Ocean Spray Cranberries, Inc.)

ders and carried half a mile or so, over a rough road, the loss from shrinkage and decay will be very considerable....

The berries were unnecessarily handled; the trouble of measuring a large lot of fruit, while the pickers were standing around, impatient to get home, was very great; and the accounts, kept under such circumstances, were not always to be depended upon. Hannah would keep her own account; and if, in the settlement, yours did not correspond with it, what could you do but allow hers?

The peck box, developed in New Jersey in the 1860s,

the six-quart measure and the tally system solved these problems. Cranberry barrels (and later boxes) were brought onto the bog and conveniently placed. The pickers dumped their measures in the barrels and received a ticket marked with the amount of that measure, for example, one peck, one bushel or six quarts. These tickets were totaled up for the final accounting.

Cape Cod grower James Webb devised a popular tally system in which each picker was assigned a numbered pail and the proprietor or tallykeeper entered the amount in the pail on the same-numbered page in an account book. Webb explained,

The book has an index of numbers upon its pages, of from one to two hundred. Every

Native Cape Codders, ca. 1890. (Courtesy of George Decas)

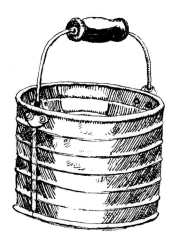

picker's name is written at the top on one of these pages, so that when a picker brings in a pail to be emptied, and calls out the number in the index, the overseer turns to the page bearing that number, and gives the credit to the proper party, thus avoiding much delay, and lessening the liability of mistakes.

A measure can be filled by a smart picker in fifteen minutes. A gang of eighty pickers can pick five barrels in fifteen minutes, or twenty barrels an hour. It [therefore] takes a person of good ability, one who is quick and expert, to attend to the duties of accountant when the cranberry picking is lively.

Using this system, the tallykeeper/overseer was not nuisanced with presenting the picker with so many tickets and writing down each quantity, although it was still the tallykeeper's responsibility to make sure that pails were filled properly.

The pails are required to be heaped up to allow for the poor berries, the stems and leaves. The pickers are usually paid ten cents per measure of six quarts; but in some places they are paid by the bushel…Each gang of fifty pickers has an overseer, whose duty it is to see that the hands do their work well, pick clean, and do not tear the vines. (Webb, 1886)

Berries being brought ashore on rail cars, ca. 1910. (Ocean Spray Cranberries, Inc.)

Below: Loading boxes of berries onto a hand-pulled wheel cart. From Country Calendar, 1906.

Opposite page, top: Knocking off for lunch, from Country Calendar, November 1906. "Oh, those picnic lunches at the nooning while backs straightened and fingers relaxed....There was plenty to satisfy the cranberry bog appetite of a hungry boy. There was the spice of good stories and old time familiar songs." From Now I Remember, the autobiography of Thornton Burgess.

Opposite page, bottom: From the shore en route to the screenhouse. From Country Calendar, 1906.

The tallykeeping system became the standard and was used on some bogs well into the 1950s, when hourly wages became universal. It was an efficient system, which probably accounted for its longevity.

Going Ashore

With so many mariners among the early cranberry growers, it is no wonder that bog vernacular was seasoned with seafaring colloquialisms. For example, as soon as one crossed over the marginal ditch to the dike surrounding the bog, one had "gone ashore."

The bags, barrels or boxes of berries filled by the pickers had to be taken ashore to be sorted and packed. In the early days, this was done by loading them onto a stretcher-like frame, or handbarrow, which two men carried to a wagon that carted the load to the sorting area. Early growers also used mules, hand-wagons, rail systems and wheelbarrows to bring berries ashore. Later, bogbuggies—rebuilt Model A's and Model T's—were used almost universally to haul harvested berries. Not surprisingly, they are still in use today.

Screening, Storage and the Separator

In the early years, it was common for the grower to dry and cool his berries for 24 hours, to allow the "field heat" to dissipate, before sorting and packing them for shipment or long-term storage. He could spread the berries out across the floor of his boghouse, cover the filled barrels with canvas or set the berries in specially designed drying racks. The berries also needed to be cleaned before shipping, which meant manually separating the good and bad berries and weeding out the "chaff," or the leaves and twigs.

One early sorting or "screening" method was to roll the berries down a smooth, shallow, inclined trough, picking out the chaff and soft berries and letting the firm ones roll into a barrel. Another method was to pour the berries slowly onto a white sheet and let the wind blow the twigs and leaves to one side.

A postcard of women sorting berries bogside on Cape Cod, ca.1900. The berries were sorted into trays and allowed to dry out in the afternoon sun before being placed in storage or packed for shipping. (Ocean Spray Cranberries, Inc.)

A watercolor of a cranberry screening crew. Unknown artist, ca. 1889. (Courtesy of George Decas)

Screening racks were probably introduced in the 1860s. Many growers devised their own racks, but a popular style emerged called a "banjo board." This was a 6-foot-long, narrowly triangular tub about 3 feet across at the wide end (called the base), with 6-inch-deep sides and a 6-to-8-inch opening at the narrow end (called the mouth). The rack was set up on two sawhorses and pitched so that the base was about a foot higher than the mouth. The bottom of the rack was a series of slats made of common lath set about a quarter-inch apart and usually running lengthwise down the rack's incline.

Once the berries were dumped in at the base, up to six people standing around the rack worked them over, allowing the debris and small berries to drop between the slats and pushing the good berries down the screen and into a barrel at the mouth. The berries that fell through the slats were collected in pans and sold as "pie" berries.

The screening rack, or "banjo board." From Corbett, Cranberry Culture, 1903.

A grower and his screeners, ca. 1890, sorting berries and pushing the good ones down the rack and into the barrel, at far right. (Massachusetts Cranberry Experiment Station)

Two drawings from Joseph White's Cranberry Culture *showing Fenwick's portable cranberry fan and winnowing separator (far right). White gives an operational description of the winnower: "The endless apron, A, forming the bottom of the hopper, gradually carries the berries forward, and drops them upon the inclined plane, B, from whence they pass to the barrel. Motion is imported to this apron by a belt connecting with the farther end of the fan axle. While the berries are passing through the air channel, C, a strong current from the blower separates the trash from the fruit." This principle is still used in separating cranberries.*

Below: At Rocky Meadows in Middleboro, ca. 1880, cranberries are sorted and packed in a makeshift screenhouse. (Ocean Spray Cranberries, Inc.)

In the 1860s, growers adapted the grain winnowing technique to cranberry sorting. Winnowing used a fan to blow the chaff and some of the bad berries away from the good berries as they were poured into a hopper. Joseph White suggested clamping a movable fan to the edge of a barrel, "somewhat in the manner that a clothes-wringer is fastened to the tub":

The idea is for the picker to pour a peck of berries into the hopper, and turn the crank while they are running through. The invention is not patented, and we give it to the public for what it is worth.

The cranberry separator, developed in the 1880s, improved on the winnowing technique. This seemingly complex machine is based on the simplicity of the bounce principle. It separates the good from the bad fruit by directing the firm, bouncing berries into one receptacle and the soft, bounceless berries into another. Several people contributed to the refinement of the cranberry separator: John Webb and James Fenwick of New Jersey; Lothrop Hayden and Hugh R. Bailey of Carver; Herbert K. Rowland and Lars H. Larsen of Washington state; Christian Frederickson of Wisconsin; and Walter Trufant of Whitman.

Eventually improved separators could grade berries by size as well as blow out the chaff and cull the bad fruit.

Grading was done by either passing berries over precisely spaced wire rods—the small berries dropping through the mesh and the larger berries rolling onto a conveyor or into a receptacle—or by running them through two oppositely revolving roller brushes that forced the soft berries to drop into a separate receptacle.

The advanced technology of the separator did not replace screeners. Berries were still screened by hand to sort them according to color and shape and to weed out any imperfect ones the separator missed. Screening racks were set up alongside the separators to receive the berries after they were winnowed or graded. Like screeners today, most early "screeners" were women.

A ripe crop was ready for shipment after screening, but growers sometimes stored the berries for up to three months in their "boghouses" to await the best market conditions. Sometimes, too, they harvested their crop unripe and let it "color" in storage. Proper storage of berries was important. Growers wanted a firm, red fruit by the time the market was ready because soft, poorly colored berries were good only for pie filling or for canning and often brought less than one-third the price.

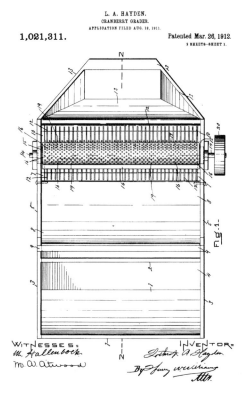

L. A. HAYDEN.
CRANBERRY GRADER.
APPLICATION FILED AUG. 18, 1911.

1,021,311.

Patented Mar. 26, 1912.
2 SHEETS—SHEET 1.

Top left and right: The cranberry separator, which marked the growers' entry into the age of mechanization. Larger bogs, higher yields and the promise of increased profits set the stage for the separator, which improved and streamlined the preparation of berries for shipping. (Ocean Spray Cranberries, Inc.)

A 1912 patent illustration of Lothrop Hayden's cranberry grader. Once in the grader, berries rolled down an inclined floor and onto a screen (12) "which is constantly swept by the fingers (19) of two aprons (16)...The berries which are thicker or larger in diameter than the distance between the bars of the screen roll down the screen and are swept into the chute (13) and thus guided to a suitable receptacle." Berries smaller than the distance between the bars immediately fell through the screen and into the "pie berry" chute.

An early brochure for the Hayden separator with attached grader claimed that its results were "truly awe inspiring and makes you feel the presence of the great inventor who first conceived the idea that sound cranberries thrown against bouncing-boards in a certain way would determine their own course in accordance with size, shape, condition and soundness..." (Courtesy of Nancy Davison)

77

One early way to speed the ripening process was to spread the fruit over a shaded area of the ground or bog-house floor. Joseph White noted that,

> This fact was well known to the enterprising inhabitants of the "Pines," who were wont to gather the natural cranberries in an unripe condition, in order to secure them before their neighbors. The white specimens thus obtained were invariably spread out and colored under an arbor of green boughs and leaves, made thick enough to exclude the sun's rays. (White, 1870)

He warned growers against trying to hasten the ripening process by putting the berries in the sun. Sunlight caused "a portion of them to decay very rapidly." In 1886 James Webb observed:

> A great many growers do not think it profitable to keep berries for any length of time, as the extra price which may be obtained does not make good the loss from shrinkage. The most extensive growers ship their crop…as soon as it is gathered.

For those growers who chose to disregard Webb, the proper storage method was to handle the berries carefully to prevent bruising, pack them in ventilated boxes and keep them at temperatures just above freezing.

In the 1860s, Joseph Hinchman of New Jersey had an innovative method for keeping cranberries in his bog-house:

His plan is to put the fruit in large shallow boxes, with perforated bottoms; these boxes are then stored in tiers, and a current of air, made alternately wet and dry, is forced up through them by means of a blower, propelled by water-power. Blowing a dry current for a few hours, and then moist air for the same length of time, it is claimed, has a tendency to prevent decay, and also to redden the light-colored berries. (White, 1870).

Barrels and Boxes

In packing, we use the "Association" barrel of one hundred quarts, made and branded by the cooper as per act of Massachsuetts legislature. The berries are shaken into the barrels, and the head pressed in place by a screw, so that the barrel is filled solid... They weigh from one hundred and fifteen to one hundred and twenty-five pounds per barrel, when ready for shipment to the railroad. (O.M. Holmes, 1883)

Inside a small screenhouse, ca. 1910. By the 1880s, screenhouses had become an absolute necessity for a successful cranberry business, large or small. (Middleborough Public Library)

Left: The screening room of Marcus L. Urann's Central Packing House, located in the South Hanson building that was to become the first home of Ocean Spray Cranberries. The berries came from the separators upstairs, through a chute in the ceiling and onto the women's screening tables. The men loaded the filled barrels onto a rail car and wheeled them to the loading dock at the other end of the building, where a railroad car awaited them. This was the most advanced screening and packing operation of its time. (Ocean Spray Cranberries, Inc.)

Opposite page, top: A Carver bog and screenhouse, ca. 1890. (Ocean Spray Cranberries, Inc.)

Opposite page, bottom left: An unknown screenhouse, 1880s, Wareham. (Courtesy of Richard Fielding)

Opposite page, right: Screening at Crane Brook in South Carver, ca. 1910. (Courtesy of Nancy Davison)

Heading-up a barrel, 1920s. Cape Cod grower Harrison Landers secures a barrel head in his backyard workshop. The final step in preparing cranberries for shipment was to pack them tight in the barrel and seal down the barrel head. This was accomplished with a "header," a screw-like vice with long iron side pieces that hooked on to the bottom of the barrel. Once in place, as explained by Webb, the header "...by the working of the screw presses the head of the barrel down in its place, where it is secured by the driving of the hoops." (Courtesy of Doug Beaton)

Right: Inside H.R. Bailey's screenhouse in South Carver, ca. 1910. Bailey, like most other small growers, did his own coopering. (Courtesy of Nancy Davison)

Opposite page, left: A shipment of empty barrels made at the E. & J.C. Barnes Mill are ready for delivery to the grower. (Ocean Spray Cranberries, Inc.)

Up until the turn of the century, each time the cranberry pickers filled their measures "crowning full," they dumped them into barrels placed strategically throughout the bog. The barrel was the cranberry's shipping and storage container for many years, and even though no longer used it is still the standard unit of measure.

In 1850 the Commonwealth of Massachusetts officially recognized the increasing importance of the cranberry industry by passing a law ordering that,

> Cranberries and all other fruits hereafter shall be measured by the strike, or level measure, that is, in the manner of flax and other small articles are measured.

Early growers made their own barrels, but it wasn't long before local coopers and carpenters seized the opportunity to service the growing cranberry industry. Joseph White wrote in 1870 that new barrels should hold two bushels and three pecks, but that second-hand barrels holding three bushels "must be filled, and then will sell for no more than the standard barrel." He reasoned that "...the peck of fruit thus lost would pay for the new package," which would encourage growers to use standard barrels rather than improvise. Standard

barrels sold for 58 cents in Philadelphia in 1868.

According to White, early New Jersey growers used barrels for shipping but preferred to load the berries into ventilated bushel boxes on the bogs because the boxes allowed for proper drying before sorting and storage. Growers in Massachusetts began using boxes on the bogs in the late 1880s, although barrels alone were used for shipping until 1900. Boxes gradually replaced barrels for shipping as well as storage over the years, until by the mid-1920s, barrels had completely disappeared from use.

To Market

Time was when the cranberry was not valued more than the common barberry. But people have lived to discover its excellent qualities, and since it is so highly appreciated for its culinary purposes, there are those who are willing to pay an almost fabulous price for the berry. It has become in many families a necessary luxury. The wealthy would as soon part with the apple as the cranberry, and it is the rage among the rich, and even those who are not so fortunate, for this fruit, which keeps it up to that price which puts it beyond the reach of the poor. (Eastwood, 1856)

Cranberry shipping containers, made at Harrison Cole's box mill. Until the end of World War I, barrels were still the preferred shipping container for berries, although a divided 40-pound box was also in some demand. After the war, the standard 100-pound barrel gave way, first to 1/2 barrels and 1/2-barrel boxes, and then to the 1/4-barrel box—easy to handle at the screenhouse and holding just enough berries to sell before they lost their freshness. By the 1930s 1/8-barrel boxes, primarily for shipment to Europe, and the novelty-size 1/16-barrel box had been introduced.

Top left: Harrison Cole's box mill, North Carver, ca. 1870. There was a working mill on the Cole homestead in Carver from the early eighteenth century until the 1950s. John Cole Jr., who built the homestead in 1710, had a grist mill that operated for 150 years. He also started a board mill whose up-and-down saw was not replaced by a circular saw until more than a century later. In 1800 Hezekiah Cole Jr. started a basket factory, importing basketmakers from England and planting a willow orchard to supply his raw materials. Harrison G. Cole took over the mill in 1843 and retooled for the manufacture of shoe shipping boxes to supply the thriving local shoe industry. In l895 the mill was again retooled, this time by Theron M. Cole for a cooper shop for cranberry barrels. First Philip Cole and then Frank Cole took over the mill, seeing it through the transition from barrels to boxes. After a fire destroyed the mill in the early 1950s, Frank Cole rebuilt and continued for a short while. But the use of cellophane bags for cranberry packaging made box making no longer profitable. The mill closed in 1955. (Courtesy of Larry Cole.)

The market value of cranberries in the eighteenth century was not quite as good as it would become by Eastwood's time. Wilmington native James Walker wrote in 1790 of two merchants from North Woburn who brought 600 bushels of cranberries to the Boston market and ended up heaving them off a city pier when no one would buy them.

By 1820, however, Boston buyers went looking for cranberries and traveled to the Cape by packet ship to conduct their business with growers at the docks. A Harwich grower might load ten barrels onto an oxcart, haul them to Brewster and drive home with $200 in gold in his pocket.

The Commission Merchants

Of such profit is the cranberry, that growers have been visited by city dealers a month or six weeks before the berry has been ready to pick...Some took them as they bid for the whole crop, and others refused. Even last season (1855), growers received from ten to fifteen dollars per barrel. (Eastwood, 1856)

One Boston merchant whose name became well-known throughout Cape Cod was Joseph H. Curtis. In 1826 Curtis was selling uncultivated cranberries grown along the Sudbury and Concord rivers at his Quincy Market stall for 50 cents a bushel. One Lincoln farmer, F.A. Hayden, received $600 for 600 bushels in 1830.

According to O.M. Holmes, Curtis began making trips to the Cape in 1847 and paid from $6 to $8 a barrel for berries packed in flour barrels and shipped to Boston by water. On one Cape journey in 1855, Curtis met a rival Quincy Market vendor named Stacy Hall and the two merchants eventually formed a partnership. "[Curtis & Hall] probably bought and sold more cranberries than any other Eastern house during this time," said Holmes in 1886. In later years, Hall joined with Horatio C. Cole to form Hall & Cole Company.

In 1857 Capt. William Crowell, a prominent Dennis grower, and George Baker of Boston established the first New York commission house. For many years Baker & Crowell was the only commission house in New York that handled Cape Cod cranberries.

There were two types of cranberry merchants: commission merchants, who received a percentage of the total retail value of the berries they sold; and dealers, who bought the berries outright and took their chances in the marketplace. Most of the Boston merchants were dealers and often resold berries to the commission merchants of New York, Philadelphia and Chicago. They were the minority, however; most growers sold their berries directly to the commission houses.

About the first of September the commission houses of New York and Philadelphia commence to mail their cards and circulars to the growers, and some of them pay personal visits to the Cape, soliciting the sale of berries. Each grower has his favorite "house" and generally continues with it year after year, more especially if the "house" is prompt in sales and remittances, and gets a good price for the berries. He takes great pride in bragging to his neighbor that such and such "house" sold his berries for $10, when his less fortunate neighbor only realized $9 for his. This is but human nature. (Holmes, 1882)

Commission merchants normally showed up on the bogs to make their offers when the crop was ready to ship. They dickered with the growers over prices, which

The Central Packing House, South Hanson, ca., 1907. Cranberry grower and entrepreneur Marcus L. Urann revolutionized the cranberry processing industry in the early 1900s by consolidating screening, grading, storing, packing and shipping under one roof. Urann's three-story, brick central processing house in Hanson, which handled berries from 600 acres of bogs, now accomplished with humidity and temperature controls what formerly required eight separately equipped boghouses and crews.

The packing house was located in the center of town, making it convenient for workers. It also had a railroad shipping dock that was ideally suited for shipping any size load, year-round. Of his facility, Urann boasted:

> The cooperage plant is in the basement of the building, where hoops from the West, staves from Maine and heads sawed at one of the Company's own mills are assembled and set up, thus the fullest economy realized. No carting or hauling of empty barrels, as is the case with some bogs, where the barrels are hauled twelve and fifteen miles.

In 1911 Urann reported that other growers were realizing the advantages of his packing house. "Yielding to requests," he was screening and packing for others at an increase of 700 percent over the previous year. Many growers' cranberries were sent all over the country from Urann's shipping dock:

> Not until 1910 were car loads shipped South, but now even the small cities order from one to four cars and Oklahoma used some 100 cars last year. (Urann, Cranberry Harvesting and Packing Under Modern Methods, 1912)

(Ocean Spray Cranberries, Inc.)

often changed daily depending on the projected harvest and how much was being picked on a given day.

> In the 1880s buyers and commissioners from the big buying cities would come to the bogs with the money that talked—cash—and the phrase "Cash on the barrel head" originated with the Cape Cod cranberry growers who never equated a promise of a dollar with the dollar itself. (Sullivan, Cranberry King)

Once the merchants and growers agreed to terms, the barrels were loaded on freight cars and shipped to the merchants' commission houses.

Exports

> There is not the slightest doubt that as the American cranberry is superior to the English or Russian, a market will be found for it, at paying prices, in almost any part of the civilized world. (Eastwood, 1856)

From 1828 to 1870, only 40 barrels of cranberries were exported to Europe from Boston, not because of a failure to find a market there, but because of the high prices fetched in America. In New York in 1868, cranberries sold for $15 a barrel, and as long as such high prices held, growers looked primarily to the American hinterland for export.

Berries shipped to England in the mid-nineteenth century were packed in small water-filled, hermetically sealed bottles. The "Cape Cod Bell Cranberries," as the English called them, sold by the pint in London for four shillings sterling. By 1868, berries headed oversees were no longer shipped in water because, as Joseph White observed,

> ...in these days of quick passages, all that is necessary is to select good keeping berries from well matted vines, and ship them in new, dry barrels, well packed, to prevent shaking and bruising.

They Picked Cranberries

Joseph D. Thomas

The Immigrant

Workers "gathering" cranberries in a picnic-like atmosphere is a popular portrait of harvesting before 1900. Unfortunately, few observers gave detailed accounts of the nature of cranberry picking or how pickers felt about their work. Joseph J. White painted this picture in 1886:

The picking season is a pleasant one to both picker and proprietor. The weather is proverbially fine…when women and children turn out in great numbers to join "cranberry picking" frolic, with well-filled dinner baskets and happy countenances…The price for

picking averages about fifty cents per bushel; the hands at this rate, making $1 per day, although a "right smart" picker can, where the berries are numerous, earn $2 per day.

In less than twenty years that picture had changed. Cecilia Perry Viera of West Wareham, remembers her picking days in the early part of this century:

I picked cranberries by the measure on my hands and knees....And your nails and all the skin comes out of your hands from the dry vines. And your knees sore....And you used to get ten cents a measure.

The cranberry pickers of Massachusetts are men, women and children who for over a century have worked not for the economic good of the nation or for the joy of frolicking on a golden autumn day. They have done it for themselves, strictly to earn a living wage. Like most laborers in America, they are often called "hands," as if they could be bought like a pair of gloves, put to work and set aside at the end of the day.

Before 1870 pickers and bog workers on Cape Cod and in Plymouth County were drawn from the family or the community; as the industry grew, people from surrounding towns (in some cases Native Americans) provided a mostly seasonal workforce. Between 1870 and 1900, growers started building larger bogs and harvesting larger crops. Labor shortages followed, and growers were forced to import laborers from the cities. Conveniently, Massachusetts at the turn of the century was bloated with a hearty labor pool.

Among these itinerant laborers were the Irish, Italians, Syrians and Slavs from Boston, Worcester and Brockton; and the Portuguese, French-Canadians and others from New Bedford, Providence and Fall River. By the late 1890s, two ethnic groups would become the foremost hirelings in cranberry work in southeastern Massachusetts—the Finns and the Cape Verdeans.

Soon after the immigrants started arriving, the family and community as a labor source dried up. It was not desirable for young children and women to work side by side with immigrant groups or men prone to curse and sweat, and locals preferred to work in year-round, higher-paying, less arduous jobs. The cranberry industry, like other industries at the turn of the century, thus became dependent on the surplus of immigrant labor,

which was cheap, docile, hard-working and, once the picking season ended, expendable.

Most of the cranberry laborers had few complaints. They enjoyed their work and were proud of their achievements. Some continued to work in the industry as it grew in the twentieth century; others moved on to other things.

The story of the cranberry workers is largely unrecorded. Only the well-manicured bogs from Dartmouth to Yarmouth, from Duxbury to Nantucket, testify to their achievements. Yet most growers realize the worth of the foremen and crews who work alongside them, those who share with them the agony of bad harvests and the sweetness of a bumper crop.

The Children

Amelia, 12 years old, cranberry picker in the third grade in school (which is in session) said, "the Superintendent (of schools) came to our Bog today and said we got one more week to pick." She picks 22 pails a day—7 cents a pail. (Lewis Hine, 1911)

One of the women pickers said "My husband is

A Syrian family from Boston, photographed by Lewis Hine in 1911. Minnie Zeadey of Wareham, who was born in Lebanon and met her husband Taft on George Briggs' bogs in Plymouth, remembers: "A group would come down from Boston every season and live in the shanties. It was like a vacation to them. In the evening they made a fire and toasted corn. Someone brought chickens from George Besse's place in East Wareham and the men would jump over the fire in dance. We sang till late at night." (Library of Congress)

Opposite page: An unknown bog in Plymouth County, ca. 1910. By the turn of the century, men, women and children from outside the region and the country descended on southeastern Massachusetts for the harvest. This migration coincided with the invention of the industry's first widely used "harvesting machine," the scoop, which radically changed the harvesting end of the industry. What had once taken gangs of a hundred to pick in a week, crews of less than fifty could now reap in two days. The scoop also affected the division of labor on the bog. For the first few decades, only men were allowed to scoop while women and children continued to handpick. Eventually, however, with the introduction of different scoop designs and the realization that men weren't necessarily the most productive pickers, everyone used the scoop. (Ocean Spray Cranberries, Inc.)

Swift's Bog in Falmouth, 1911. Photographer Lewis Hine captioned this photograph: "Carrie Medieros ready to pick. Said twelve years old—second year picking. The manager said 'We have 150 workers besides the kids'" Manny Costa of New Bedford remembers working as a child: "We missed 6 weeks of school.... We just took off [from New York] and settled in. The laws weren't that rigid." (Library of Congress)

Below: For this picture, also taken in September 1911, Hine wrote: "Near Parker Mills. Three pickers going home from work. Anne Benoit, 7 years old, her brother Vincent, 11, and sister Inez, said 6 years old, 'and picked last year wid me mudder.' Smallest one, not quite large enough to work. Father works in Parker Mills." (Library of Congress)

dead. I got two children. I got to make 'em work, you can't take 'em to school yet." (Hine)

Children had picked cranberries alongside their parents for as long as cranberries had been harvested, but by the turn of the twentieth century, child labor on the bogs had taken on new importance. Now migrant families who traveled long distances required lodging for the picking season, and children had become part of the working family unit. The survival of the family, it was felt, depended on their contribution.

Several factors promoted child labor. Cranberries were big business, families needed money and the country was bulging with an ever-increasing population of immigrants scurrying to fill a host of menial jobs. Since hand-picking was still being used and growers were paying by the measure, children were not considered a burden. Extra hands were needed to harvest the berries before the frost set in.

Reform groups such as the National Child Labor Committee claimed that long hours of work hurt the spirit of the young by depriving them of an education and a normal adolescence. And many people believed that child labor kept adult wages low and thus weakened the nation by promoting an underclass of dispirited, disgruntled workers who would view their jobs as worthless and never strive for anything more.

> When the Jersey harvest time approaches, the crowded tenement districts of Philadelphia literally disgorge their denizens to the cranberry bogs. Last autumn on six bogs 864 children, ranging in age from four to fourteen years, were found at work. Of this number 603 were ten years of age or under. (*Survey* magazine, 1911)

> The testimony of school superintendents and principals confirms the judgement of reason, that when these children return two months late for the school year, it is with deadened faculties and jaded nerves. (*Survey*, 1911)

> Young children were found working long hours under a padrone, in Massachusetts as in New Jersey, and families were crowded into unsanitary shacks. However, as a much greater proportion of the harvesting in Mas-

sachusetts is done by scoops operated largely by Portuguese men, the evils of child and woman labor were found to be not as extensive as in New Jersey, where nearly all picking is done by hand. (*Survey*, 1912)

Child labor laws in Massachusetts prohibiting the hiring of children under fourteen were largely ignored, but at least they were in place. In New Jersey, where children were critical to the industry, child labor laws were virtually nonexistent before 1920. The *New York Times* reported in 1916:

Infantile paralysis quarantines have accomplished more in one season to bring about the elimination of child labor on the cranberry bogs than all the years of agitation by social workers. Ordinarily, more than 5,000 pickers are required to gather the harvest… In the past years it has been a big event among these foreigners [Italians] to make their annual pilgrimage, children and all, and…pack the family purse. It is estimated that in the past at least 1,200 children have helped….Because of the general demand for labor in all of the industries, the gangs of adult pickers who have come to the bogs, leaving the women and children behind,

have shown an independent spirit. Most of them have demanded higher wages than they got last year or the years previous.

Child labor continued in both states well into the Depression years, when the huge surplus of adult labor finally destroyed the competitive edge of hiring children, and workers demanded its prohibition. After this time, children still accompanied their parents to the bogs, sometimes helping to carry empty boxes, but their picking days were over.

Bog Shanties

They keep this up, going from bog to bog in the neighborhood, and picking while the season lasts. [They]…camp out in their tents, or erect rude dwellings, and live in them until the season is over. (Webb)

By the 1890s, pickers were coming from as far as a hundred miles away to the bogs of southeastern Massachusetts. Whole families joined groups of up to 150 to 200 pickers to "take the picking by contract." During this time, and for many years to come, the cranberry industry required thousands of workers for the harvest, which lasted from late August to late October. To keep workers for the entire picking season, growers had to provide them with a place to stay.

Belford Coldas, 8 years old, September 1911. Hine wrote of Belford: "…lives in New Bedford, picking on Week's Bog near Waquoit in Falmouth. Fifty workers—ten children from 7 to 10. The boss's boy earned from $1.00 to $1.50 per day. We bought a scoop from the boss who said that they made some of them smaller for the children's use. A Portuguese father near Week's Bog said he made $6 a day himself and several others picking in his family. Said that lots of children pick after school begins." (Library of Congress)

Left: "Group of workers at Smart's Bog near South Carver. Annette Roy, 7, the youngest worker, picked last year. Napoleon Ruel, said 9 years old. Both live in Fall River. Bosses are posted every 20 or 30 yards keeping the people at the picking line. The manager, a veritable slave driver, was an old sea captain who threatened the workers, 'or you'll go ashore'.…. The boss of one large bog said they used to use women and children there, but now they use scoopers entirely and that is work for men. Scoops are faster." (Photograph and caption by Lewis Hine, September 1911; Library of Congress)

An engraving of a boghouse from *Cape Cod Cranberries*, by James Webb.

Below: "Shack near John D. Crocker's Bog, housing seven Portuguese in bunks.... There were bunks for twelve persons. The shack was 10 feet x 12 feet and 6 feet high.... The boss said that his workers are in the factories the rest of the year and give up their jobs for a time to come down here and get a holiday. He goes every year to get them and says, 'We don't pay them till the end of the season or they would all leave anytime.'" (Photograph and caption by Lewis Hine, September 1911; Library of Congress)

James Webb, in his 1886 booklet *Cape Cod Cranberries*, described the "Bog-House" as a building used for housing pickers and storing berries. The size and number of boghouses depended on the size of the bog. For instance, a 10- or 12-acre bog required a boghouse of about 18 by 30 feet:

> The Bog-House...will accommodate thirty-two hands, or a sufficient number to take care of the bog in the picking season. The lower floor is used as a cook room and as a room in which to dry the berries. The upper floor is arranged with sleeping accommodations. (1886)

By the early 1900s, the storage functions of the boghouse were being taken over by central processing facilities. Boghouses became primarily workers' shelters, commonly called "shanties."

In a 1914 report by the U.S. Immigration Commission on Seasonal Agricultural Laborers, Professor Alexander A. Cance observed that proper housing in labor-intensive, seasonal agriculture was a problem. He credited the cranberry industry with doing a better job of providing housing than did the construction industry, probably because of the presence of women and children pickers. Nevertheless, he reported that sometimes men lived "in abandoned houses and sheds near the bogs, in great filth and squalor."

> On some bogs small two-story houses, usually 10 by 12 feet, are built to accommodate the men or the families. The first floor is used as a kitchen, and the upper floor has bunks built of rough lumber. On other bogs, acting on the theory that anything will do for six weeks, the owners provide shacks of the roughest and crudest sort, and no attention is paid to sanitary arrangements or cleanliness. (Commission report, 1914)

> Certain medical and hygienic authorities declare with conviction that the exposure to rain, cold and malarial atmospheres are provocative of fevers and tuberculosis, and that neither the water supply nor the unhygienic surroundings are conducive to physical well-being. (Commission report, 1914)

Cance pointed out that the squalid living conditions on some bogs affected more people than just the pickers. As the pickers came into contact with people from town, diseases they carried arising from an unsanitary environment could be easily spread.

Overcrowding was common in the workers' shanties. In 1911 the National Child Labor Committee visited bogs on the Cape to determine if the state of Massachusetts was allowing violations of its child labor laws. They reported that in Falmouth, "one bog owner (Mrs. Phinney) told us that 21 workers on another bog live in one small house and she thinks they must sleep standing up."

Manny Costa of New Bedford came to Wareham with his father and two brothers from New York in the 1930s to work on Makepeace's Frogfoot Bog. There were several shelters at Frogfoot, including the bosses' shanty and a women's shanty, that altogether housed about 40 people. Manny was 13 when he first made the eight-hour trip from New York in his brother's reconstructed Model A jalopy.

> Our shanty had 10 men. The one room on the bottom was a kitchen, with no sink, no

water, a stove and a table. Upstairs was two long bunk beds where the 10 of us slept, two windows. Talk about poor air circulation. With the residue heat that stayed there, you'd practically die from it. The windows were nailed shut. You didn't want them open because of the mosquitoes. We used to have a mattress cover that we filled with straw. We'd rake the straw up outside. This would be our bedding.

Although we lived under very crowded conditions, I don't remember ever minding it. It really didn't bother me, except that I had to sleep between my father and older brother on a straw mattress. My little brother slept beneath the bunk. Other than that the inconveniences weren't that bad because we were used to inconveniences.

The shanty was just for sleeping and cooking. The kitchen area was so crowded and hot because of the wood stove, especially in mid-August, that you couldn't stay there. When the food was cooked, we'd sit outside on the ground on logs.

Father did most of the cooking. The 10 of us made a verbal contract that we'd share the food expenses. At the end of the season, when we got paid, we'd split it up.

It was a great time. It was a festive time. When cranberry time came, everybody looked forward to it. We knew there was work and money to be made. It was family. You ate well that time of year. There was always food.

Bottom left: Hine's photo of the inside of a New Jersey bog shanty: "Shanty dwelling of cranberry pickers, settlement called Rome, on Whitesbog, Browns Mills, 1910." (Library of Congress)

An elderly picker for the Ellis D. Atwood Company, taking tea on an autumn day off, courtesy of foggy weather. The characteristics of bog shanties were fairly consistent: a pot-belly wood stove, bunk beds or cots, a hand pump outside, no sink. The calendar dates this picture to October 1953. (Photograph by Ted Polumbaum)

89

The Finns

Linda Donaghy

The Finns call their homeland *Suomi*, the Finnish word for swamp, because much of Finland is swampy, as well as heavily wooded and dotted with thousands of lakes. The swampy woodlands of Cape Cod reminded immigrant Finns of *Suomi*, and many eventually settled there out of nostalgia for what they had left behind. Fortunate for them there was work to be had, much of it on the Cape's cranberry bogs.

The Finns' migration "from across," in the late 1800s was, like so many immigrants', for economic betterment

A group of Finnish cranberry men in the late 1920s. Walter Heleen is second from right. (Courtesy of Paul and Linda Rinta)

Below: Finnish women from Worcester at the United Cape Cod Cranberry Company's Indian Head Bog in Hanson, ca. 1924. (Courtesy of Wilho Harju)

Opposite page: A family portrait, 1920. Finnish families often teamed up to work the bogs. This group includes (left to right): Mr. Kari, Tyyni Kari, Mrs. Kari, unknown, Olga Niemi, Toiva Kari, Karl Niemi, Gertrude Niemi and Kusti Lehto. (Courtesy of Mary Korpinen)

and escape from political oppression. The first to arrive in Massachusetts came to Quincy to work in stonecutting and shipyards; to Worcester and Fitchburg to work in steel mills and nail factories; and to Fall River to work in textile mills. According to Wilho Lampi,

> They left Finland because there was nothing there. There were two classes—peasants and the upper classes. If you were a peasant, you were a peasant. You had to work for the landlord and you never owned anything.

The first Finns on the Cape probably came about 1885 and settled around West Barnstable. According to John Saarimaki, among those early arrivals were Gustav and Aukust Silverberk, Erik Erikson, John Maki and Tuomas and Andy Kaski. Some worked as farmhands at first and some as woodcutters and carpenters. Others found work in A.D. Makepeace's West Barnstable Brick Company. By the mid-1880s Makepeace had built the "Big Bog" at Wankinco, and his Frogfoot bog was under construction. He needed workers to build and maintain these big projects, and his Finnish brickyard workers provided a ready source of labor.

Four women, knees protected by padding, ready for picking, ca. 1934. Left to right: sisters Ellen Maki and Gertrude Rinne, their aunt Katri Harju and friend Saima Waino. (Courtesy of Gertrude Rinne)

Right: Andrew Palm working the bog under the supervision of his son-in-law Otto Salmi (background) around 1920. For Andrew working the bog was a holiday compared to the Pennsylvania steel mills he had left behind. (Courtesy of Mary Korpinen)

News of work to be had on the bogs spread, and by 1900 other Finns were coming, settling in the Tihonet section of Wareham, East Carver, West Wareham and Middleboro, as well as Barnstable, Sandwich and other Cape towns. This was when George Eastman, Alex Johnson, Fillus Harju, Thomas Suominen, Verner Kumpunen and many others arrived. Grower Wilho Harju recalls the story of his father Fillus Harju's journey:

He came to Worcester in 1902. There was no future in Finland for a young man. The eldest son inherited the family property and the others would have to work for him for peanuts. The Russians were drafting Finns to serve in the army, so many of them left to avoid the draft. He got a job at American Steel & Wire Company and worked there seven years for 72 hours a week. He worked inside this mill drawing wire to make nails, running it through vats of acidic compounds. The doctor told him to get outside or he would die from the place.

When Fillus heard about the cranberry bogs, he and a friend left Worcester for the Hatchville section of Falmouth. He worked for a while in bog construction and brought his family down as soon as he found a place to live. He also picked cranberries, farmed and did odd jobs until, in 1910, he was able to buy about five and a half acres of land with a house and a bog. Then he grew cranberries and had cows and a vegetable garden. He also worked building bogs for others.

One of the first Finns to own bogs on a large scale was Alex Johnson, who came from Finland to Maine in 1897. He heard about the cranberry business through friends and came to work at the Federal Cranberry Company in South Carver. He and George Eastman, another Finn, worked together building bogs. Alex was a foreman for Federal, but after fifteen years he decided to strike out on his own. He bought a house with four acres of land and began building his own bogs.

In the early years of the twentieth century, Johnson was the largest Finnish cranberry grower in southeastern Massachusetts. His son Carl recalls:

There were eleven kids in our family. Our house and land came to be known as Johnson village. During the Depression at harvest time a gang would come from Fitchburg and Gardner, Finnish people out of

work. We had a camp at that bog for the workers and some stayed in other places. My father took a loan out for that property from a local bank. We kept a big strawberry bed on that bed. We sold them to the Decas brothers. That paid the interest on the land. As kids we would head toward the Cape in a Model T touring car to pick blueberries. That used to help pay some of the bills.

Before Alex died he sold much of his land and built homes and bogs for his children. Today Carl Johnson owns and manages some of those bogs. He also managed cranberry bogs for others and installed plastic for bog sprinklers.

Carver growers Charlie Kallio and Edwin and Mary Korpinen remember their parents' stories of immigration. Kallio's parents came from the mills in Fitchburg to the Cape to work. Like so many others, they had heard about the cranberry industry from a friend:

> The majority of the immigrants were single men. They were housed for free in bog shacks with no plumbing, only an outhouse, with three to four men sharing a room. The shacks were segregated by nationality. My father lived in these shacks learning the industry and saving his money. The women came later and also worked the bogs. (Charlie Kallio)

> My father was a farmer in Finland but he couldn't make it there. He first settled in the north Woods timber country. Mary's father worked in the steel and coal mills in Pennsylvania. Other people came at harvest time from Fall River and New Bedford to pick cranberries. The Finnish people saw something to be made here, and they had the insight to foresee a good life in the cranberry industry. (Edward Korpinen)

Many of the early Finns worked in bog building, weeding and ditching as well as picking and sorting. In reminiscing about their life in the cranberry industry, many Finns speak of *sisu*, a word meaning great inner strength and fortitude. It was *sisu* that helped the Finns endure the difficult times.

Onni Ahlberg, using a snap scoop on Clayton McFarlin's bog in South Carver, 1938.

> My mother would pick blueberries and strawberries in the summer to save money. Winter came and you tightened your belt. There was not much work. We cut wood for heating, but work was scarce. Everyone had animals and raised their own vegetables. You only went to the market for flour, salt and coffee. We had two cows, one or two pigs and thirty to forty chickens, plenty of eggs. There were seven kids. My mother worked all day weeding— weeding and screening were the women's job. My father was a foreman on a bog. The first piece of land he bought was three acres of bogs. I would work them every day after school. It was a family effort. (Charlie Kallio)

Mary Korpinen's father, Otto Salmi, came from Finland in 1911. He worked in steel mills and lumber camps across America. He and her mother, Anna, who came here as a child in 1904, settled in Carver. Anna worked in screenhouses with her mother and sister.

Three Finnish teenagers, William Ahlberg, Sigrid Ahlberg and Reino Harju, ready for the bogs with their working gear in hand—snap scoops and six-quart pails, ca. 1930. (Courtesy of Wilho Harju)

Right: (Left to right) Reino Harju, Ellen Maki, Fillus Harju and Gertrude (Harju) Rinne on a Carver bog, ca. 1940. (Courtesy of Gertrude Rinne)

My mother spoke of walking from [her] Huckleberry home to the Federal Cranberry Company to screen cranberries, a distance of about eight miles, and earning ten cents hourly. My father worked on Mr. Waters's bog in South Carver as a foreman, plus he had six acres at home to tend. He sold milk and eggs from the animals we kept for extra money. My mother worked the bogs all day, too. They came home to tend the garden. She grew vegetables, potatoes and fruits. She worked the bogs till she was in her seventies. She was a strong woman.

Hard work paid off for the Finns. Early on they were in demand as foremen and supervisors. According to Marsha Penti, a writer on the early Finnish experience in America, at one time almost all of the bogs had Finnish foreman, and it was a not uncommon belief that if a Finnish foreman left a bog for another one or to work his own, the bog he left would suffer.

John Saarimaki was born in Michigan in 1905 but grew up in Finland. He returned to America in 1930. He tells the story of how he became a foreman for the Makepeace Company:

I was working on the bog one day when I was told by one of the foreman, Frank Butler, that they wanted me to replace the other foreman who had joined the service. I'd only been in this country for a few years and I couldn't understand why he wanted me to be foreman. I told him I'd think about it. I didn't want to be a foreman. A couple of days later he called over to me, told me to come with him and told me my new job would be to oversee another bog. He was making me a foreman whether I wanted to or not. I didn't even have the chance to give him my answer.

Although there were some Cape Verdean foremen, the Finns were given more opportunity to advance. Nonetheless, the two immigrant groups got on well together as a rule, occasionally socializing on and off the bogs during picking season. According to one Cape Verdean,

The Finnish were always there. They were friends. The two groups were pretty much on the same plane. So many were such good

friends and good neighbors. And we lived like a family.

By the 1920s enough Finnish families had settled in connected areas of Middleboro, Carver and Wareham to have established an active community. According to Penti, Carver had 44 Finnish households of 200 persons; Middleboro had 16 households of 80 persons; and Wareham had 29 households of 136 persons. West Wareham was the center of the community, and along Main Street the village was known as Finntown, home to the Zion Evangelical Congregational Church (there was another church, *Soini*, in South Wareham), the Socialist Workingmen's Hall and the *Amerikan Suomalainen kansanvallan Liitto*, the Finnish dance hall.

The church and the hall helped the Finns maintain their cultural identity, but they also divided the community. Many Finns who went to the hall for Saturday night dances never made it to the church on Sunday morning, and many of those who regularly attended church never stepped foot in the hall. Vieno Saarimaki's family had strong church ties. She recalls never going to the dance hall or dancing at all, because "It was forbidden by my parents." Yet her husband John spent many a night dancing at the hall. In spite of disapproval, the hall was an important vehicle for Finnish ethnicity, promoted through dance, drama, song and traditional celebrations.

The Finns kept coming to the cranberry regions from Finland as well as from other parts of New England in the early years of the twentieth century. As a group, the Finnish immigrants had a 96 percent literacy rate, which helped them assimilate and, with their love of the land, helped them understand the economic value of owning property and managing their own enterprise. They were eager to acquire land, and they got their opportunity to do so before and during the Great Depression, when land was cheap. Some brought property with their savings or borrowed money from successful Finnish bog owners such as Alex Johnson. Others, like Ed Korpinen's father, borrowed money from Marcus Urann.

The Finns quickly proved themselves capable farmers. An Economic Recovery Act report issued in 1934 showed that of the several hundred cranberry bog owners in Massachusetts, approximately 70 were Finnish. Most of them owned 10 or fewer acres.

At the turn of the century, A.D. Makepeace bought up the old iron foundry holdings in Tihonet Village and

A group of Finnish workers on United Cape Cod Cranberry Company bogs in Plymouth, ca. 1948. (Ocean Spray Cranberries, Inc.)

converted the factory buildings, houses, canals, reservoirs and swamplands to serve his cranberry empire. Many Finns who worked on Makepeace's bogs lived in Tihonet Village. Vieno Saramaaki remembers:

The people there were very close. We were all in similar situations. People were nice to one another, we never locked our doors. The Tihonet Village had three duplexes alike in a row and one big house on the hill for two families. We initially had an outhouse. Later we had the indoor plumbing, running cold water.

It was hard at times but we got by. My husband [John Saarimaki] and I met at a screening for Wankinco. I was nineteen and staying with people in Sandwich. I needed a ride home so he took me. That started it all. I picked as a teenager. Then I planted vines. That's when all the bogs were being built. It was cold, wet work. We would dress in layers of clothes and boots or rubbers. We worked eight hours, it was mostly women. This was

artist, still makes these rugs.

Saunas are very popular in Finland. Almost every farm has a bathhouse, and most towns have public saunas. The bathers sit on wooden benches and throw water over hot stones to produce steam. They then rinse in a cool shower, producing an effect of complete physical and mental relaxation. In the United States the Finns built their bathhouses away from the main house, and preparing the sauna was an all-day affair. Mary Korpinen has vivid memories of relatives and neighbors sharing this Finnish custom:

> Every Finn had a sauna. Once a week you'd heat it up to have a bath. In those days we didn't have bathtubs. It would take my mother all day to prepare it. There was a big fireplace to heat the stones. It was like a big ritual, and around 2 o'clock it would be ready. Others would come and use the sauna. My mother would make coffee and a marble cake, and my sisters and I would eat all the nuts off it. The Hayeses would come and the Griffiths, my grandparents and the Kangas with their boys. It was like a party. These are very fond memories.

Today the Finns are considered to have a knack for cranberry growing. They have developed cultural practices, such as the use of fertilizer, emulated by most successful growers. Larry Cole, a Carver grower, says that, "…as far as growing cranberries is concerned, they're probably the best."

Many of the inventions that achieved wide use in the industry over the years were created or improved on by Finns. For example, the Mathewson picker was developed in 1923 by Oscar Tervo, who was born in Finland and came to live in Quincy, Massachusetts. And the snap scoop, first developed in the nineteenth century by Donald Lumbert, was refined by a Finn named Kataja. Everett Niemi, a mechanic who worked for the A.D. Makepeace Company, developed a method for making and packing dehydrated cranberries during the war. According to John Saarimaki, Niemi "could work with anything."

Perhaps the success of the Finns as cranberry growers can be attributed more than anything else to *sisu*—the fortitude that helped them never to lose their desire to work the land. Even in difficult times, the value they

Oscar Tervo, of Quincy, at the controls of the Mathewson Picker, which he invented in 1920. (National Archives)

Opposite page, left: Elsie Salmi and her friend Frances Fernandes, working the Salmi bog, ca. 1940. The Salmi's 5-acre bog provided a little extra cash for the family while father Otto worked as foreman on Benjamin Waters' bogs. The two teenagers shown here were neighbors and close friends; they worked after school and on weekends for money to buy their school clothes. (Courtesy of Mary Korpinen)

Opposite page, right: Alex Johnson and Cecilia Fernandes. Cecilia, born in 1890, came from Cape Verde to California in 1916, and then to South Carver in 1920 with her husband Peter. She worked on local bogs, including Alex Johnson's, and raised produce on her own land. During the Depression Peter returned to California to work, while Cecilia stayed in Carver, eventually buying some small bog acreage near her home. Cecilia's daughter Frances is in the picture to the left. The Finns and the Cape Verdeans had much in common. They came as migrant laborers and worked side by side. Many lasting friendships, such as with the Salmis and the Fernandes, were established. (Ocean Spray Cranberries, Inc.)

in the 1940s. From there I worked in a screening house on Main street in Wareham. I liked it there; it was near my home and I liked my work. Some couldn't do the work, they would get too dizzy. I got used to it and enjoyed the other women I worked with. We would sing or talk together. I really missed my work when I left. This life gets in your blood. I really enjoyed it.

Many Finns lived in Tihonet. Erik Erickson bought swampland and an old farmhouse at the "Apple Farm" across from what is now the Carver landfill. A schoolhouse that was once housing for the iron workers is where the Ericksons and many other Finnish children went to school.

Traditions from Finland found their way into the cranberry life. For example, making rag rugs is a modern version of the old Finnish craft of making *rijijy*, a rug used for sleigh covers. It was common for women in Mary Korpinen's neighborhood to share the looms used in rug making, and she remembers sharing a loom at Carl Johnson's house across the street. Her daughter, an

placed on land never diminished. Following the cranberry scare of 1959, when others were bailing out, many Finns seized the opportunity to expand their holdings. Edward Korpinen remembers:

> My dad kept buying more land. He could see a future in cranberries. During that time we sold our cranberries for peanuts. I went to work in a textile mill for a few years just to be able to pay the taxes on my property. I didn't want to lose it. I hated the mill, but I'm glad I held onto my land.

Today many third- and fourth-generation Finns are still growing cranberries, and the tradition of passing down land continues. The Finns have made their mark on the cranberry industry of Massachusetts. The industry owes a great deal of its success to the people of Suomi.

The Cape Verdeans

Marilyn Halter

…if at a holiday dinner we think of any one group of people as having contributed most of the cranberry part of our repast, the Cape Verder is the one to have in mind. Over and over again, and without contradiction, owners and overseers of cranberry bogs pronounce the Cape Verder, whether he picks by hand, scoop or snap, the very best harvester of cranberries and spreader of sand with the wheelbarrow on the Cape Cod bogs. (Anthropologist Albert Jenks, 1924)

For 500 years, until independence in 1975, the Cape Verde Islands off the west coast of Africa were under Portuguese colonial rule. During that time, drought, natural disasters and colonial exploitation resulted in a legacy of famine and high mortality for Cape Verdeans, many of whom responded by emigrating to the United States.

This search for a better life began in the late eighteenth century with the arrival of single men who had been picked up by whaling vessels and ended up on American shores. By 1900 there was a pressing need for cheap labor in both the expanding textile mills and the cranberry industry of southeastern New England, leading to a steady influx of Cape Verdean immigrants to this area.

Cape Verdeans were known as rugged farmers, and they brought a robust aptitude and passion for the land to their work on the cranberry bogs. Yet few Cape Verdeans became owners of cranberry bogs, remaining migrant laborers and residing off-season primarily in New Bedford and Providence. In early September, they drifted back to the cranberry district. One New Bedford reporter in 1900 said of their return:

> By the end of the summer…They have folded their tents like the Arabs and silently stole away to Cape Cod's cranberry bogs.

Some Cape Verdeans managed to buy or build cranberry bogs and thus realized a portion of the American dream. Others saw cranberry picking only as a chapter in their lives that brought back pleasant memories of bonfires and dewy mornings, or of storytelling and record-breaking scooping. Still others remembered picking for its backbreaking toil for low pay.

As the late Antonio Jesus, former whaler and cranberry foreman for the Fuller-Hammond Cranberry Company, described it:

> The hardest job I've run across is picking cranberries. Of course, construction is hard but there's money in it so you don't feel it. But you take a man, go out and wheel sand nine hours for $1.80, 20 cents an hour, and you have got to do it. If you stand around, if you don't put a big load, the boss says, "What's the matter? You going to travel? You left a place for a suitcase?"

For Cecilia Perry Viera of West Wareham, the feeling is much the same:

> I can look at cranberries, yes, but not eat them. It was so hard on your hands. It tore the skin off and got under your fingernails. And it hurt your knees to kneel there in the bogs for so long.

Photographer Lewis Hine, who documented child labor in the region in 1911, recorded the following conversation overheard at Hollow Brook Bog in Wareham:

> A tousled headed boy about 10 years old comes up with a box of cranberries balanced on his head, struggling with both hands to keep it up. The checker is very profane. (The boss on the field is worse.) "Put it in there, god damn it. Hold on, god damn it, go back and fill it up." (The box is heaping but not quite

heaping enough.) The checker has told me that the boxes are supposed to hold two measures (12 quarts) but really hold 13 1/2 quarts. "There ain't no need of cursing" someone says. "Well, I ain't cursing, god damn it, but go back and fill it up." A little boy of 12 is picking vines from the barrels and hears the checker say, "Take the vines out. Throw 'em to hell overboard." The checker keeps up a running fire of this "speeding-up" sweatshop talk. He calls across the field to a boy half-way in with his box. "Go back and fill it up or god damn it, you'll go home." (Library of Congress Archives)

All of the hardships characteristic of migrant labor were experienced by the cranberry pickers. Yet in comparison to factory work, to congested city life, to unemployment and employment discrimination, the weeks of the cranberry harvest were a welcome change for many. And the wages they could earn during a good season would be sufficient to take them through the winter with extra to send back to the old country or, in some cases, to bring other family members to the United States.

The pickers were frugal. A count made in 1908 at one bank in the cranberry district showed five hundred Cape Verdean depositors with savings averaging from two to

A Cape Verdean crew on the bog, ca. 1910. Once the dew dried, scoopers began their day's work. (Ocean Spray Cranberries, Inc.)

Opposite page: A young girl scooping on an Ellis D. Atwood bog, 1938. Many pickers rented their scoops for the season from the bog owner. In the late 1930s A.D. Makepeace Company charged about $20 to buy a scoop and $3 to rent it. According to Manny Costa of New Bedford, who with his father and brother picked for Makepeace, "If you snagged your scoop into a stump and it broke off some teeth, then you'd have to have it repaired and you would've lost time, and time meant money." (Middleborough Public Library)

Cecelia Perry Viera, 11 years old, working at Eldredge Bog in Rochester, 1911. (Lewis Hine photograph, Library of Congress)

the Crioulo version of rice and beans. Some Cape Verdean cooks used the large wooden mortar and pestle, or *pilon,* imported from the islands to grind corn in the traditional manner.

Picking cranberries was our main thing for the winter to pay bills and do what-so-ever. Buy your food by the sack and put it in the house, sugar by the hundred pound, butter by the five-pound tub, rice, beans. We had plenty of beans from the garden. We had our meat from the pig and chickens and eggs. We made our own home butter sometimes. We lived. (Flora Monteiro)

Nobody knew how to make the cranberry juice. That's a fairly new product and so are the pastries. My mother used to make the sauce and jelly and we used to have it on all the holidays. All the owners allowed us to have a few cranberries if we wanted them. Either that or we'd handpick the underberries. (Albertina Fernandes)

I remember my wife used to say, "Bring home a couple of quarts of cranberries today" and I would bring them home from the bogs and she would make something good. (Antonio Jesus)

three hundred dollars. Some accounts held as much as fifteen hundred. At the end of the cranberry harvest, the bank paid out over twenty thousand dollars in savings to Cape Verdeans. Workers were paid once at the end of the season and lived on credit in the meantime.

On the bogs, single men stayed in shacks or "shanties," as they were called—crude, shed-like structures, often overcrowded with barely enough room to stretch out on straw or sleep on flimsy mattresses. Workers were expected to gather their own deadwood for heat and cooking and provide their own food and utensils.

Some owners of more or less isolated cranberry bogs provided cottages for foremen near the bogs. Increasingly, though, Cape Verdeans who had the savings and had let go of the notion of returning to the islands bought cheap lots and erected cottages, or bought land for small farms. The immigrants who moved to the Cape and Plymouth County as year-round residents were able to recreate their traditional habits more freely in a rural setting.

Many of the oral histories describe the family garden, canning and preserving the produce and preparing *Crioulo* recipes from home-grown food. Most popular of the staple dishes were *manchupa,* or *cachupa,* a hearty stew made with samp (a coarse corn meal) and cooked with potatoes, squash and linguica or other meat, and *jagacida,*

Often school-age children did not start the school year until the end of the cranberry season in mid-October. Some towns issued permits allowing parents to keep their children out of school until the harvest was over.

In the oral history, *A Portuguese Colonial in America,* educator Belmira Nunes Lopes reminisces about picking cranberries as a young girl in Harwich and Wareham:

During my high school years, I don't believe I ever started school at the beginning of the year in September...because my parents were itinerant agricultural workers. They picked cranberries in the fall, and sometimes, when the cranberry season was over in one bog, it might still be going on in another, and they'd move from one bog to another to help pick cranberries, screen them, and box them.

For the children, cranberry harvesting did have its

100

special moments. Lucillia Lima reminisces about the songs her grandmother sang while picking:

> I liked to be with my grandmother when she picked. The work was so hard, but she talked to us and told jokes. She'd sing old Cape Verdean songs, the one about the sea of the full moon, rolling on the beach and playing in the sand…I remember the phrase in one: "What killed him was the tongue of the world." Grandpa Cy always sang "Bye Bye Blackbird," wherever he went.

Lucillia also remembers the delights of storytelling in the evening hours during the cranberry season:

> We always looked forward to Saturday nights. The neighbors' children would all come over and we would have hot bread with cranberry jam and hot cocoa…Yho Lalla was a great story teller. He was much older and had a white mustache and smoked a pipe. He told us stories of Yho Lobo, the wolf. Yho Lobo was always the bad guy and some other animal was always the goodie two-shoes. The stories recounted all the terrible things that happened to the wolf, how he fell into the hot fire or had nothing to eat in winter. The moral was to mind mother and never be lazy.

Many recall the socializing after a day of picking:

> On the cranberry bogs, there was storytelling at night, playing the guitar, music, singing, dancing, parties. We always had a party after picking time. We had a square dance and they would call the dances in Crioulo. Contra dances we call it. (Lucillia Lima)

> After christenings, we had a dance and we'd serve *canja* at night, a thick chicken soup, crackers, and anything else my mother could prepare, and we'd dance from night until morning. Even the young children…would be allowed to stay up as late as they could stay awake. (Belmira Nunes Lopes)

> We always had a good-bye party at the end of the season. Some went back to Wareham,

Maria Gamboa (Lucillia Lima's grandmother), picking on an Ellis D. Atwood bog, 1938. Lucillia remembers her grandmother as a very liberated lady,

> *She was a very big woman—six feet. At first she picked by hand, by the measure, and the men would pick with the scoops and would make more money. So she said she wanted to pick with the scoop and the men didn't go for it because she picked very well, more than many of them . . . so they had to put her in her own section. And she'd be way ahead, often finishing her section before they finished theirs. She'd tie her head and she'd sing her Crioulo numbers, her Cape Verdean songs, and she'd push ahead. (Middleborough Public Library)*

Left: Day care, ca. 1918. (Ocean Spray Cranberries, Inc.)

101

some to Providence, others to Taunton. We'd all chip in and bring something to the shanty. (Mary Barboza)

Many Cape Verdean pickers were proud of their scooping abilities. For women in particular, the stress of working at a "piecework" rate, combined with the pressure of competing for this privilege with men, prodded them to become superior harvesters.

We went picking one Sunday to this place. My nephews went and when it was 2 o'clock, I asked the fella, "How many boxes have I got?" He say, "72." I say, "Lord, 72! I'm going to make it 100," and I grabbed my scoop and boxes on both sides. I went over and found a plot and when I got through with that plot, I had 24 more boxes. Lord, I thought, I'm going to make it to 100. I made it to 100 and said, "That's it!" (Flora Monteiro)

When I picked, I picked by the box. I did piece work. A lot of the women didn't want to do piece work. I never liked hourly work. To me, it was too slow. I could make three times more picking by the box than I could by just the hourly rate. With the hourly rate, you go on your knees and scooped one scoop at a time. I just couldn't pick at that pace... The most I can remember picking in one day is 88 boxes. There's supposed to be a picture of me and my brother—my brother is quite the picker—taken on a day that I picked 100 boxes. I don't remember that—100 boxes! (Albertina Fernandes)

I picked cranberries for Makepeace Company. One time, the boss says, "Flora, I want you to come pick cranberries over here to a section that I got here. I don't want to put all the crowd in there. I want you alone. Come with me." That place was just like you took a bucket of cranberries and emptied it down on the ground. I didn't see him but he told me afterwards that while I was picking, he took out his watch and timed me. He says, "Flora, every minute you had a box. You were filling those boxes so fast that I had to time you." (Flora Monteiro)

Besides harvesting the cranberries, many women hand-set vines in the spring and weeded the bogs during the growing season. However, year-round employment on the bogs was reserved for men. They did the bog building, planting, sanding, weeding, packing, shipping and so forth, and, of course, only men were made foremen.

Tony Jesus recalled how he had to persuade his wife to move to the country so that he could accept a year-round position as a bog foreman:

In 1919, I drove team. When we got through picking in October, it got kind of chilly one

A family affair, ca. 1950. Before World War II women made up one-fourth to one-third of the picking force. They continued to pick after marrying, bringing their little ones with them to the bogs. Women and children were said to handpick more rapidly than men, and strong-handed women were believed to be the best snappers. Supposedly, however, only men were good scoopers. It wasn't until the 1930s that women were given the opportunity to scoop. They quickly proved to be at least as efficient as the men. (Ocean Spray Cranberries, Inc.)

day. I got through and I told Mr. Hammond, "Mr. Hammond, you want to get ready to pay because I've got to go back to my winter job." "Oh, no, no, no," he said, "I'm going to make you foreman in Carver because the foreman there is going back to Finland and not coming back again." Course I was young and I liked that. So I asked my wife. She didn't want to. She said, "I don't want to live in the woods." So all right, we pack up and we go to New Bedford. I went back to the mill, a cotton mill. Then I beg. I beg until she decides it's all right. In 1920, I call the old man first week in April. He said, "Come right along." I became a year-round man. I put in about 50 years in the Carver bog.

As Cape Verdeans began to settle in Plymouth County and on the Cape, they encountered increasing prejudice in these traditional Yankee strongholds. They were immigrants with unmistakable African as well as Portuguese roots and, as such, were targets of racism. For example, the Cape Verdean neighborhood in Onset was called "Jungle Town," and in September 1906 the New Bedford *Morning Mercury* carried this account of the arrival of the schooner *Flor de Cabo Verde* from Fogo Island:

The news has spread in the islands that there is work to be had near this city and everybody in the islands who can get together the price of a passage and enough money to show on arrival at this port is coming this fall. While the owners of the cranberry bogs and others interested in the property wish to see the immigrants come to pick the cranberries, it is said they do not like to see them remain on the Cape, and the problem is how to make the pickers all go away from the places where they pick berries.

As early as 1905, the town of Marion issued a directive against further employment of Cape Verdeans in public works. And in that same year, a Wareham High School senior's commencement address, entitled "Drifting Backwards," deplored the influx of "half-blood" Portuguese to the area, bemoaning the fact that "our poor American girls are obliged to labor side by side with these half-civilized blacks." Several years later, in 1917, Belmira Nunes Lopes gave a valedictory speech on the "The Ideal

Pickers on Ellis D. Atwood Bogs, South Carver, ca. 1953. (Ted Polumbaum photograph, Ocean Spray Cranberries, Inc.)

Right: A little boy carrying field boxes, 1953. When children weren't picking, especially in post-war years when child labor had essentially disappeared, they still came to the bogs with their parents. Many kept busy, like the boy in this picture, bringing the pickers empty field boxes. This small chore was very important to the "piece rate" scooper, who often spent a considerable amount of time walking to and from the shore retrieving boxes. (Ted Polumbaum photograph)

An elderly man pushing a scoop on a Makepeace bog in Wareham, 1938. Harvesting cranberries has never been restricted to the young. Older men and women, out of necessity or desire, have worked the bogs since the industry began. (Middleborough Public Library)

Right: Like the peasants on the islands, many Cape Verdean women in southeastern Massachusetts retained the custom of wrapping their hair in scarves and carrying heavy loads on their heads. It was not uncommon to see a female picker hoisting a crate of cranberries on her head to bring to the tallykeeper for a count. In this photograph, Mary Barros of Middleboro gathers berries from Homer Gibbs' Middleboro bog. Mrs. Barros picked here for many years—in addition to planting and weeding in the spring—and with her husband and three children lived in a bog shanty on the Gibbs property until she was able to build a home in Carver. (Standard-Times photograph, 1953)

Town," in which she observed that the perfect community would have no prejudice.

Many Cape Verdean newcomers never purchased land for bogs because they did not see themselves as permanent settlers. If there was extra money, it was used to buy land in the islands, not here. Converting swamplands to productive bogs takes from three to five years, and most of the immigrants simply could not afford to invest time and money in a project with such a delayed financial return.

The best cedar swamp land at that time was only $5 an acre. Then they would build the bog themselves. This was all hand labor. There were no bulldozers or anything else at that time. Yet that same $5 would be sent back to the old country. The Finnish people worked on the cranberry bogs like we did, but eventually they ended up buying and building bogs. And now in Carver they are a large majority of the good cranberry growers. Anybody could have bought bogs at the time. Cape Verdeans didn't have their minds set on living in this country. (Albertina Fernandes)

That's the biggest mistake I made. I had a chance to buy a bog. I have a friend now in Osterville who has three-quarters of an acre. He works it himself. He's got a nice home. So, you see, today the cranberry business is wonderful. Which is a mistake I made. I didn't want to have to worry about anything when I got to be an old man. (Antonio Jesus)

A few individuals did buy land and build bog. In 1925 Dr. Henry J. Franklin, director of the Cranberry Experiment Station in East Wareham, noted that Cape Verdeans owned no more than a total of 100 acres but that this was at least a start in the right direction.

Some immigrants, like the Pina brothers of Falmouth, bought homes in the cranberry region and built small bogs to supplement their income. Antonio Canto Barboza came to the Cape in 1906 and set up a store where berry truckers waited for their loads from nearby bogs. At various times, he also ran a gas pump, an ice cream parlor, a restaurant, a barber shop and a pool hall. In addition to these ventures, he and his family had a small cranberry bog and strawberry fields. In Wareham another enterprising Fogo Islander set up a store and a dance hall near the bogs, while also keeping a small piece of land to grow his own cranberries.

Those who managed to buy land and turn it into productive bogs are part of the success story of this immigrant group. Brought up as peasants on the islands, the Cape Verdeans sustained their connection to the land and at the same time achieved economic security. This is no small feat.

Albertina Fernandes, from the island of Fogo, and her son Domingos are long-standing cranberry bog owners in South Carver. She describes the quality of her life:

We just enjoy it out there. It's beautiful to walk around the bog. You see all kinds of animals—deer, muskrat, rabbits, fox. And it's yours. You go up and walk around and it makes you feel free and airy. It's just a beautiful feeling. All during the time the kids were growing up, we used to go camping out there. Like early in the spring, my husband would come home and say, "Oh, let's take a ride." We'd go up there and walk around the bogs. It was just nice, relaxed, no stress. No matter what you have to do, you're not under complete stress all the time. The boys are all college graduates and they still love to go out there. They really don't have to do it. They enjoy it like I do.

Cape Verdean women and men scooping, ca. 1948. (Ocean Spray Cranberries, Inc.)

The Labor Strike of 1933

Marilyn Halter

In part because of the seasonal nature of their work, it was difficult for the cranberry pickers to organize. But that didn't mean they were happy with the status quo. As early as 1900 a riot broke out at a bog in North Carver, where worker dissatisfaction with low wages led to fist fighting and arson.

By the Depression years, a more organized effort took place. In September of 1933, 1500 workers went on strike demanding an increase in wages, guaranteed employment until the end of the season and recognition of their right to unionize. The first order of business was the establishment of following wage scale to be followed by the growers:

Men scooping:	.80/hour
Women scooping:	.70/hour
Handpicking:	.15/measure
Women screening:	.40/hour
Men weeding/sanding:	.50/hr; 40 hr/wk.

The union members also demanded the immediate formation of a workers' committee to routinely inspect living and working conditions on the bogs, and a shift to payment at two-week intervals rather than once at the end of the season, with credit slips acceptable at the local stores and banks issued between paychecks. Prohibitions against children laboring on the bogs was also a demand.

Workers claimed that some owners were unfairly paying by the box when berries were sparse, and by the hour when the crop was abundant. One picker said that he was paid 30 cents for harvesting 32 quarts when only a few others were on the bogs, which was cut to 18 cents when more laborers showed up. Allegations were also made that some workers were illegally fired for joining the union. According to James Bento, a Cape Verdean judge who was an attorney at the time, the strike was initially organized by outsiders, but many pickers, after a generation of poor working conditions, readily took part in hopes of improving their situation. Others supported the strike out of solidarity or fear of reprisal.

The work stoppage spread to some 40 bogs throughout the Carver, Onset and Wareham region. Strikers demonstrated their growing militancy by donning arm bands with the word "picket" and driving trucks with signs such as "We Want A Living Wage" through the area.

The cranberry growers responded to the strike by posting land with "no trespassing" signs. They also deputized approximately one hundred foremen "and other reliable whites" from Plymouth and Wareham to "protect the workers" and "drive agitators out of the cranberry bogs."

> At that time they had hired cops because of strikes going on at different bogs and they had the hired men dressed as cops. The hired man said, "What do you think you are?" I said, "I know what I am. I'm a cranberry picker for Makepeace, but I want what I picked marked down. I ain't making the boss no present." And the cop came to me and

Pickers at the Smalley bog in Wareham, at work during the strike of September 1933. In retrospect, many growers believe that the pickers' monetary demands were not unreasonable. One grower said, "The pickers were entitled to more. It was the only time of the year they could get it [money]—to get a little extra for the winter. Harvesting rates probably should've been higher than other jobs. It was harder work. They looked forward to picking season, just as we did." (Standard-Times photo)

says, "Lady, you got to keep quiet or else I'll take ya." I said, "You won't dare put your hands on me." I called him all kind of names. After the season was over, the hired cops became the workers and they were working right along with my husband. So I told my husband. "Where did he leave his uniform, in the sand bank?" (Flora Monteiro)

On September 14 violence erupted. A local bog owner and Carver Selectman shot a Cape Verdean worker, wounding him in the hand. The worker had allegedly tried to wrest a gun from the owner and the gun went off. State police quickly arrived on the scene. They fired shots that wounded several people and arrested 64 pickers for assault or attempted assault. The Wareham selectmen filed a request for martial law, and word came from Boston that Governor Ely was watching the situation closely.

The most complete account of the strike comes from the files of the Wareham *Courier*, which consistently downplayed the purpose of the strike, thus making it difficult to reconstruct what really happened. This much is clear: Outside workers were brought in by the growers to pick the already rotting berries and the union organizers were arrested on charges of falsely representing themselves to the workers in their signature campaign. They were convicted and given two-month sentences.

Perhaps the most significant factor in dissipating the momentum of the strike was several days of heavy rains during the third week of September, at the near height of the harvesting season. With the bogs flooded by the ceaseless downpour, picking had to come to a halt, regardless of the organized work stoppages. There were no benefits for striking laborers, which meant no harvest wages to carry them through the winter. At this point, the union drive began to dwindle, and the strikers, hungry and desperate, gradually went back to work under police surveillance.

Although strike efforts did not result in a general wage increase or unionization, in some cases settlements were made for higher wages on individual bogs. More important, this was the first strike by agricultural workers in the history of Massachusetts, as well as the only labor dispute that involved primarily Cape Verdean immigrants.

Examples of individual as well as organized acts of protest against working conditions abound in the oral histories. One outspoken young woman stood up to the

tallykeeper when he tried to cheat her:

One time I picked four boxes in a half hour. That time you had to take them to shore. And I was right near the edge so all I did was just jump and put them on the shore. And the tallykeeper says, "You picked four boxes right now? What do you think you are, a machine?" He didn't put the four boxes down. He put three down just to be mean. And I was tired. I opened my lunchbox. I couldn't eat because I was so hot and sweaty and tired. And I said, "You better put four boxes down." So Tosy, the boss, heard me arguing with the tallykeeper. He says to the tallykeeper, "If she told you she picked four, she picked four. Whatever she tells you. I know her long enough." (Flora Monteiro)

Tallykeeper Clara Tamagini, checking the measures of Mary Cruz, Mary Barros and Bella Barros, September 1933. Most workers continued to pick after the first few days of the strike. Many admitted that though they supported the strike in principal, they couldn't afford to hold out for their demands. (Standard-Times photograph)

Tools of the Trade

Joseph D. Thomas

Since the early nineteenth century, cranberry growers have produced nearly all of their own tools and machinery—from the most primitive hand tools, such as the dibble, the turfing axe and the harvesting rake, to the most sophisticated power-driven machines, such as wa-ter-reel harvesters and sanding barges. The first tools used in the cranberry industry were those adapted from other farming practices, reshaped by growers to fit their needs.

Today many growers have converted their old screen-

The Norway scoop, 1860s. This scoop, used in Scandinavia to harvest blueberries, may well have been the model for the cranberry scoop. (Ocean Spray Cranberries, Inc.)

Right: A horse-drawn vine-setter, ca. 1900. (Ocean Spray Cranberries, Inc.)

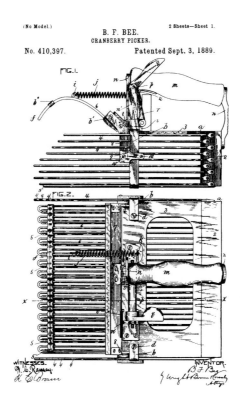

(No Model.) 2 Sheets—Sheet 1.
B. F. BEE.
CRANBERRY PICKER.
No. 410,397. Patented Sept. 3, 1889.

houses into storage sheds, garages or machine shops for maintaining their equipment. Here one can find state-of-the-art technology alongside idle antiquity. One thing is evident everywhere: Little comes from the assembly line of John Deere or International Harvester. Gary Western, manager of the Federal Furnace Cranberry Company, explains why:

> We make almost all of our equipment. I just finished making this water-reel harvester. We don't patent anything. Hell, everyone makes

their own anyway... You take ideas from here or there, but you make it the way it suits your purpose. We've been making all of our equipment for years. I've designed a lot of it, but we all work on the same ideas and techniques... until someone else comes up with a better one. For example, this water-reel—they started this in Wisconsin, but we changed it by adding the ditch reel. You won't find anything on these bogs made by the big companies. They wouldn't bother. Christ, if John Deere or one of those companies tried to make machinery for the cranberry industry, they wouldn't sell more than 50 of a kind. So we make our own.

The Snap Machine

The first major breakthrough in the growers' pursuit of a harvesting machine that would eliminate their dependency on a large labor force, and improve productivity, was Daniel Lumbert's invention of the "snap machine" or snap scoop in 1887. The Lumbert snap consisted of a

A harvesting machine from the early 1900s. This machine, pushed along the bog like a lawn mower, had levers attached to the handle bar. When squeezed, the levers pulled two wires controlling the teeth on the harvesting box, causing the teeth to trap the vines against the box and tear the berries loose. (Ocean Spray Cranberries, Inc.)

Top left: The "snap machine" or "snap scoop." (Engraving from L.H. Bailey, Standard Cyclopedia of Horticulture, 1928)

An engraving of the Bee snap machine, patented by Benjamin Bee of Harwich in 1889. Bee's machine worked on much the same principle as the Lumbert snap, but in his patent application Bee boasted of certain improvements "whereby the strength, durability, and efficiency of the apparatus are increased...." He assigned his patent rights to Emulous Small, A.D. Makepeace and George F. Baker, three cranberry men known for their business acumen. Ironically, however, the Bee snap disappeared almost immediately. There are no pictures of it in use, and in a 1942 note to John C. Makepeace about the Bee, William Makepeace asked, "Were any of these made up? I never saw one." He wondered how much the assignees paid for the patent rights. (Courtesy of Nancy Davison)

Picking with a long-handled scoop on a Makepeace bog in Wareham, 1938. This device was the invention of Arthur Atwood of South Carver. Not very many of these were made or even seen because they were slow. As one former scooper pointed out, "You couldn't make any money with it." (Ocean Spray Cranberries, Inc.)

109

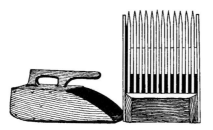

metal box with 6-inch-long teeth extending from the bottom of the front side. The top and sides were hinged with a movable front, held in place by a spring. Snap scooping was a one-handed operation; the spring was worked by the thumb and the handle was held in the fingers. The teeth were shoved into the vines and the front was sprung to them, pulling the berries into the box as the machine was withdrawn.

The snap scoop didn't break off as many vines as did the larger wooden scoop, because the teeth were retracted after they grabbed the berries. It was used primarily on younger bogs, whose vines had not become thick and matted. On the denser vines, the snap was difficult to operate because of the thumb and wrist movement needed to operate the mechanism.

The Scoop

First introduced by individual growers in the 1890s, the scoop appears to be a descendent of the Maine and Norwegian blueberry scoops. According to Joseph White, a cranberry scoop was used in New Jersey as early as the 1850s, when the wild Jersey bogs were regarded as public property. This scoop, which he neither described nor illustrated, "combed the berries off, and also pulled out large quantities of old vines and dead grasses," which resulted in higher-yielding bogs.

The most common scoops had round wooden teeth 12 inches long and a half-inch in diameter. They were set far enough apart to comb through the vines but close enough together to trap the berries.

Working from the shore inward, scoopers crouched on their hands and knees with one hand on the top handle and the other on the bottom handle. They plunged the scoop into the vines just below the fruit and pushed along the nap of the bog in a rocking motion so that the berries fell back into the scoop's box.

When demand for the scoop heightened, several manufacturers patented and began producing them: A.D.

Makepeace Company made the Whaler and Banner scoops; R.A. Everson of Hanson called theirs the Cape Cod Cranberry Picker; and Hayden, Bailey, Rowley, Buckingham and others all had their versions.

Ricky Kiernan, a foreman for the Makepeace Company who made hundreds of scoops, says that the Whaler was the most popular in Massachusetts. It was a two-handled rocking scoop with maple teeth, spruce sides and white poplar handles.

Large cranberry companies that employed picking gangs assigned scoops to pickers every morning and collected them at the end of the day. Pickers paid to rent the scoops and were usually responsible for their maintenance. The scoops were numbered and branded with the company name. Often someone was hired full time to keep the scoops in working order.

The practicality of the cranberry scoop for years eliminated the need for a mechanical harvester. As long as labor was affordable and reliable, the scoop was satisfactory, as it was efficient and did minimal damage to the vine. However, since scoop-picking was labor intensive, growers began to feel too dependent on the unstable conditions of the workplace—labor relations, wage demands and labor shortages were a nuisance, especially during harvest time. The ideal solution was a machine that would eliminate the large workforce and harvest the crop cleanly and quickly. While growers in the western states were turning more toward wet-harvesting, Massachusetts growers, convinced that the dry-harvested fresh-fruit crop would continue to be their bread and butter, experimented with various dry-harvesting machines throughout the first half of the 1900s.

Bill Stillman, foreman for the United Cape Cod Cranberry Company, overseeing a picking gang on one of the company's South Hanson bogs, ca. 1940. (Ocean Spray Cranberries. Inc.)

Scoop-picking on an Ellis D. Atwood bog in South Carver, 1937. (Middleborough Public Library)

An unknown gang of scoopers, ca. 1955. (Ocean Spray Cranberries, Inc.)

Bottom right: Exchanging a box of berries for a tally ticket, 1953. (Ted Polumbaum)

A woman putting on hockey pads to protect her knees from the rough vines. (Ocean Spray Cranberries, Inc.)

"Nearly 150 German prisoners of war assisted in harvesting the crop this season," according to the caption accompanying this picture in the New Bedford Standard-Times, October 21, 1945. (AP photo)

A healthy crop of berries waiting to be taken ashore. (Ocean Spray Cranberries, Inc.)

Carver grower George Pass, in October 1946 trying out a vacuum harvesting machine invented by Acushnet grower/inventor Frank C. Crandon, who may best be remembered for some of his innovative, moderately successful machines. This machine combined "combing" (notice the scoop at the end of the hose) and suction. However, its kinks were never sufficiently worked out to make it practical. (Standard-Times photograph)

Top right: The Brooklyn Bridge. A Washington state invention, this bridge-like contraption spanned over 150 feet of cranberry bog. The span was driven across the bog by two trucks on opposite shores, and the work crew was cradled along the span. Its usefulness was limited to bogs with parallel shores, something rarely seen in Massachusetts. (Ocean Spray Cranberries, Inc.)

A suction picker in Averdeen, Washington, ca. 1940s. The inventor, William Hoyt, and his family show off an hour's picking—four boxes on the ground and one in the tray under the hopper. (Ocean Spray Cranberries, Inc.)

114

Industry leaders looking over the Crandon picker at the Manomet Bog in Plymouth during a trial run in 1945. From left to right: Marcus L. Urann, Henry J. Franklin, Ferris Waite, Frank Crandon, Herbert Leonard, R.C. Everson and Olin Sinclair. The Crandon picker was too big and awkward, had poor picking action and was difficult to mount and drive. (Massachusetts Cranberry Experiment Station)

Mathewson pickers, Wareham 1936. This was the first successful mechanical cranberry picker, developed in 1920 by Oscar Tervo of Quincy. The Mathewson had a 6-wheeled chassis and 14 rows of picking fingers, 41 to a row. The curved fingers, attached to a revolving drum, reached down into the vines with a combing action, scooped the berries and emptied them into a chute. The berries rolled down the chute into boxes on the side of the machine. An assistant walked alongside the machine, replacing boxes as they were filled. The Mathewson was the only mechanical picker used until the 1940s, but it had a difficult time maneuvering on the small, angular bogs of Massachusetts. Also, the bog edges had to be hand-scooped first to clear the way for the machine. (Middleborough Public Library)

Three views of the Darlington picker. (Above photographs, ca. 1958, courtesy Ocean Spray Cranberries, Inc. Right: Standard-Times photograph of Hilda Pou, September 1969)

"My grandfather (Joseph J. White) handpicked, and when the first World War came along he couldn't get enough people to handpick so he started scooping," said Tommy Darlington of New Jersey. "The snap scoop was slow and expensive to make, but easier on the vines. The dry pickers we developed were like a continuous motion snap scoop." Like the snap scoop, the Darlington's pruning action left the vines in a healthier state after the picking was done. Massachusetts growers began using the Darlington in the early '50s and still use them today. Robert St. Jacques of Wareham is now the manufacturer of this machine.

Western pickers, 1958. Mrs Ernest Goddard (left photograph) leads her crew (above) on the Goddard bogs. (Ocean Spray Cranberries, Inc.)

R. J. Hillstrom, a manufacturer of picking machines for Oregon growers, brought his "Western" picker east in 1947 to try it out on Massachusetts bogs, which he found more uneven, with shorter vines, than Oregon bogs. He returned to Coos Bay to work out the bugs and was back for the 1948 Bay State harvest. Two of Hillstrom's redesigned machines went from bog to bog on a demonstration tour. The machine was so successful he had all the orders he could handle before he returned home. In recent years the Western picker has given way to Darlingtons and Furfords, which are lighter and easier to handle. (Ocean Spray Cranberries, Inc.)

The Furford picker, 1989, Harju bog, Carver. Made in Washington, the Furford was introduced to Massachusetts growers in the 1960s by Oiva Hannulla. Unfortunately for Oiva, the machine did not make a good impression at its trial run and growers turned away. One grower observed that it tended to tear up the vines and was hard to maneuver. With some refinements, the Furford showed promise and is now used on more bogs than the Darlington. For many growers, the Furford's sturdiness is more important than the Darlington's ease of handling. (Beverly Conley photograph)

Going ashore, LeBaron Barker's Century Bog in Bournedale, 1938. The wheelbarrow was a great improvement over horse-drawn wagons, hand-carts and handbarrows for bringing berries ashore. This iron- and wood-framed two-wheeled barrow was designed more for hauling the large loads than for ease of handling. (Middleborough Public Library)

Right, top and bottom: Going ashore, 1950s. H.R. Bailey's rubber-tired barrows were a popular means of taking berries to the shore. They were built to carry one or two boxes over the wheel for better balance. (Ocean Spray Cranberries, Inc.)

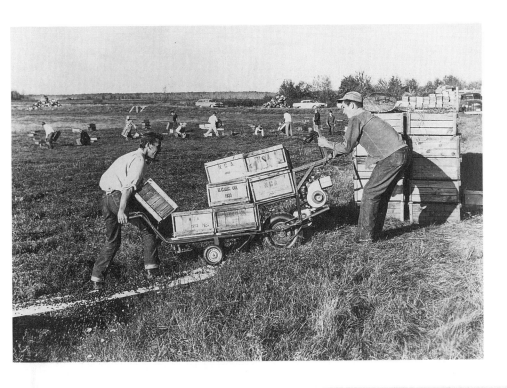

A bog buggy taking berries ashore, ca. 1950. The "bog buggy" has proven to be one the most durable and versatile cranberry industry innovations. For going ashore, a rebuilt Model A Ford, its rear end converted into a flat bed, made the perfect vehicle. Larger buggies, made from Ford or Dodge 3/4 ton chassis, are used for spreading sand or fertilizer as well as for hauling berries and anything else. Christopher Makepeace began building buggies for the A.D. Makepeace Company in 1971 and has built 28 for the company since then. He says that only trucks built between 1948 and 1955 are used in bog buggy manufacture. (Massachusetts Cranberry Experiment Station)

Top left: A mechanical wheelbarrow, ca. 1960. An improved mechanized wheelbarrow was supposed to provide relief from backbreaking manual barrow work. Instead it proved to be cumbersome, unsteady and somewhat of a flop. Notice in this photograph the cranberry boxes tipping over into the ditch. (Ocean Spray Cranberries, Inc.)

Loading filled boxes from buggy to truck for shipment to the screenhouse. (Ocean Spray Cranberries, Inc.)

Winnowing berries on the shore, 1950s. Bags of berries harvested with Western pickers are winnowed and transferred into boxes before being trucked to the screenhouse. (Ocean Spray Cranberries, Inc.)

Top right: Rows of unassembled cranberry boxes outside the box mill at Tihonet, A.D. Makepeace Company, 1950s. (Photograph by Ricky Kiernan)

Harvesting at the Wankinco bogs in South Carver, October 1960. (Standard-Times photograph)

Taking floaters with "float cradles" in the 1940s. After the harvest, the bog was flooded and loose berries left behind by the pickers floated to the surface. Some ingenious float-harvesters attempted to loosen berries from the vine by creating vibrations with power boats. (Ocean Spray Cranberries, Inc.)

Taking floaters at an Ellis D. Atwood bog, 1941. (Ocean Spray Cranberries, Inc.)

Dave Eldredge trying out a Getsinger water-harvesting machine in the early 1960s. The water-reel knocks the berries from the vine in a whipping action as it plows across the bog. The Getsinger is a Wisconsin machine that funnels the berries into a scow, attached to the rear. When filled, the scow is detached, and its contents dumped into a truck for transporting to a receiving station.

David Mann of Buzzards Bay, probably the first Massachusetts grower to water-harvest on a large scale. Mann purchased a Getsinger in 1960 but encountered difficulty from the start. Shallow areas in the unlevel bogs and bog ditches prevented him from bringing the tubs to the shore. After visiting the Western bogs, and with much experimentation, he built the area's first riding water reel. Similar to the one pictured here (below), it had three reels and angled bars for ease in removing grass caught up in reels.

David Mann's water harvesting methods in the mid-'60s induced the president of Ocean Spray, Ed Gelsthorpe, to rework the plans for the company's Middleboro facility to include a receiving station for water-harvested berries. It was a good move. Today about 65 percent of all Massachusetts berries are water-harvested. Meanwhile, David Mann continues to experiment with new machines and cultural practices, leading the way in water-harvesting innovation in Massachusetts. (Top photograph, Massachusetts Cranberry Experiment Station; bottom photograph, Ocean Spray Cranberries, Inc.)

David Mann's adaptation of a backhoe to lift a tub of berries onto the truck, 1962. The truck was also rigged with a portable drier that blew hot air on the berries as they passed over a screen and onto a conveyor. Berries destined for the freezer needed to be dried first. Mann admits, "You can never get Massachusetts berries dry. Even after they're dried, wet-harvested berries always seem to stay wet." (Massachusetts Cranberry Experiment Station)

Hog Wallow Bog, Chatworth, New Jersey, ca. 1960. Massachusetts water-harvesters visited Bill Haines' bog and were impressed by the Jerseymen's method of taking berries off the flooded bog using a conveyor or elevator. While growers in Wisconsin use a backhoe to dump harvested berries from the scows into the truck, most Eastern growers corral the berries with wooden booms toward one end of the bog, where an elevator takes them in the truck. (Ocean Spray Cranberries, Inc.)

Ready to Market

Joseph D. Thomas

The greatest improvement in the processing and handling of berries in the early twentieth century came about in 1907, when many growers joined a new marketing cooperative called the New England Cranberry Sales Company. In order to handle the huge pools of berries its members delivered for processing, the sales company

built four screening and shipping facilities. These were located in Barnstable, Plymouth, North Carver and Tremont. This was convenient for small growers, who were now spared the expense of a screening operation. Larger growers, such as Ellis D. Atwood and L.B.R. Barker, had their own screenhouses and preferred to handle berries themselves.

The semi-centralized system of handling and screening berries had its difficult side. At sales company headquarters in Middleboro, agents Susan Pitman and Arthur Benson had to make discriminating decisions as to whose berries were to be shipped where. Often they had to match a particular variety with the demand of a certain market. And when there was a surplus, berries were put in frozen storage and growers sometimes had to wait a year or more to be paid for their harvest.

For the most part, the sales company was a blessing to the region's cranberry growers, and most of them speak highly of its accomplishments. No longer were growers beaten down by greedy merchants whose visits to the region were filled with collusion and deceit. One grower remembered how the commission merchants "would meet in an apartment in Wareham. Then, one of them would come to your place and give you a very low price. You'd refuse to sell. Then another would come the next day and offer you a lower price. That's how they got what they wanted."

While the sales company helped stifle the commission merchants, it did not eliminate competition. Independent growers and private distributors effectively competed with the sales company to clean, pack and market the cranberry. Eventually, a cooperative named Ocean Spray would overpower them all.

LeBaron R. Barker's Century screenhouse in Bournedale, 1937. Built in the mid-1930s, it was considered a model packing house, and reflected the care Barker took in maintaining his 256 acres of cranberry bogs. In 1901 he bought the Crowell Brothers Bog, one of Plymouth's oldest, rebuilt it and called it "Century," for the new century.

The tower on the building was built to provide ventilation for the berries in storage and also as a look-out for forest fires. The screenhouse had an entrance for trucks so that berries could be unloaded indoors to protect them against rain. Separators were equipped with chutes that directed vines into an underground cement bin protected by fire sprinklers. Vines were bailed and hauled away—no trash was allowed to pile up outside.

Barker was considerate of his workers. His screen room had two screeners at each of the 14 screens and two inspectors, which made the work less intense. It had a separate hot water heater for temperature control. The storage shed had its own heater. On the bog, his crews were doubled up, for their own safety as well as for better bog management.

Forever the romantic, Barker told Cranberries magazine in 1943, "When I was a young man, we picked with snaps, turned our separators by hand, screened over slats, shipped in barrels, and hauled our berries over sandy roads with horses. When we watched the frost, the old kerosene lantern was our best friend, as we could tuck it under the carriage blanket when we were cold." Barker never stopped loving the cranberry industry. He died an untimely death, an apparent suicide, shortly after selling the Century Bog and his other holdings to the A.D. Makepeace Company in 1960. (Middleborough Public Library)

The no longer used Federal Furnace Cranberry Company screenhouse in South Carver, 1989. (Joseph D. Thomas photograph)

Opposite page: Berries en route to the screenhouse, via Main Street in Wareham, ca. 1928. (Ocean Spray Cranberries, Inc.)

Unloading freshly harvested berries at the New England Cranberry Sales Company's Tremont screenhouse, 1935. (Middleborough Public Library)

Top right: Unloading berries at Ellis D. Atwood's screenhouse in South Carver, 1938. (Middleborough Public Library)

Melville Beaton of the J.J. Beaton Distributing Company, gazing at his storage shed full of berries in November 1959, just after the aminotriazole announcement. At the time of this photograph, Beaton's Distributing Agency was the largest independent packer in Massachusetts. (Life magazine photograph by Ted Polumbaum)

Bailey's Cranberry Screening Units, 1940s. This assembly, which separated and screened the berries, comprised 6 units: from right to left, blower, elevator, separator and grader, double belt screen, conveyor, and box shaker. Each unit was driven by its own motor. (Courtesy of Nancy Davison)

Left: H.R. Bailey, riding a duster, and his assistant, Neil Murray, at his blacksmith shop, ca. 1940. Hugh R. Bailey (1872–1958) emigrated from Nova Scotia to South Carver (with a short stay in West Wareham), where he established the H.R. Bailey Company in 1895. Originally a blacksmith and carriage shop, the company shifted to the manufacture of cranberry equipment as Hugh himself became more interested in growing, eventually owning 45 acres of bog. An obituary says of him that "when confronted with a task he would sit down and study it to make improvements." Certainly the tasks he faced as a grower were the impetus for many of his inventions and improvements, among them separators, screens, blowers, wheelbarrows, dusters, sanding machines, box shakers and turfing and pruning tools. (Courtesy of Nancy Davison)

Workers dumping field boxes, just unloaded, into the blowers (or winnowers) of their separating machines, L.B.R. Barker screenhouse, South Carver, 1938. The berries are passed through the blowers, which winnow out the chaff. Then they pass up the elevator, are dumped into the separator/grader and move along the conveyor, which carries them into the screening room. There screeners continue the sorting or "screening" process. (Middleborough Public Library)

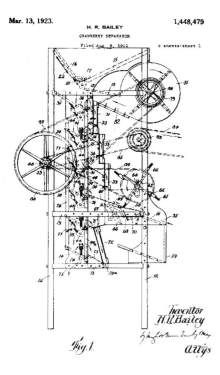

Mar. 13, 1923.

H. R. BAILEY

CRANBERRY SEPARATOR

Filed Aug. 8, 1922

1,448,479

2 sheets-sheet 1

The patent diagram of H.R. Bailey's Cranberry Separator.

Top Right: Dumping berries into a separator at the Tremont screenhouse, 1935. (Middleborough Public Library)

Bottom right: Ellis D. Atwood, inspecting berries that have passed over the bounce boards and are on their way to the screening room. The separator also grades berries according to determined sizes. (Ocean Spray Cranberries, Inc.)

Bounce boards on the back end of a Bailey separator, showing the berries as they fall and ricochet away from the angled slats. The soft berries, rather than bounce, will hit the slats and drop straight down into their own receptacle. (Ocean Spray Cranberries, Inc.)

Women screening at the Ocean Spray plant in Onset, 1959. (Life magazine photograph by Ted Polumbaum.)

Top left: A mother and son screening berries at the Tremont screenhouse, ca. 1950s. Screening berries simply means removing the unwanted ones. Berries that are not colored properly, too small or imperfect in some way are taken out one by one by the screeners. It is a tedious job that demands complete concentration. The women sit in one position for a long period, focusing on thousands of tiny moving objects less than three feet away. A conscientious employer would have excellent lighting and comfortable seating, and allow screeners frequent breaks to relieve eye and back strain. (Ocean Spray Cranberries, Inc.)

Screening at Ellis D. Atwood's South Carver screenhouse, 1938. (Middleborough Public Library)

Filling boxes at Tremont, 1936. After screening the berries travel to the packing room where they pour into boxes set on "box shakers." The shakers use a vibrating action to remove space in the box so that the berries can be packed solid. Just as cranberry barrels had to be packed absolutely tight, so too did boxes. A "slack pack" meant a considerable depreciation would be allowed at the customer or dealer's end. In this photograph the boxes are given the Honker Brand label. One of the most widely used Eatmor labels, Honker Brand was applied to at least 50 percent of all Eatmor shipments. Honker berries are Late Howes, at least 85 percent colored, with a count not over 120 per cup and fit for 20 days' travel. (Middleborough Public Library)

Top: Nailing up boxes at the Tremont screenhouse, 1935. After the boxes were shaken and the slack removed, they were nailed up, ready for shipment. (Middleborough Public Library)

Top left and right: Berries being unloaded at Quincy Market in Boston in the '30s and '40s. (Ocean Spray Cranberries, Inc. and A.D. Makepeace Co. photographs)

A rail shipment of berries leaving Tremont in 1936. One of the inconveniences of the New England Sales Company's Tremont screenhouse was the absence of a rail spur leading to the loading dock. Though the main rail line was only about 1000 feet from the screenhouse, the adjacent property owner (a small cranberry grower) refused to let the cooperative tie into it by cutting through his land. As a result, the berries had to be trucked to the train. (Middleborough Public Library)

Cranberry labels came into use in Massachusetts in the late nineteenth century. Early labels were used by commission houses and distributors from New York and Boston simply as a way to identify their product.

Around 1912 New England Cranberry Sales Company growers, under the direction of their parent group, the American Cranberry Exchange, began using the brand name Eatmor to classify their berries. Under this system, berries were graded with almost scientific precision. The percent of coloration, keeping-quality, size and variety were all factored together to determine the brand. For example, Skipper Brand was Early Blacks, averaging 75 percent colored with not more than 10 percent white berries; a count not over 125 to 150 per cup and fit for 15 days' travel. Blue Bird was a mixed variety, two-thirds colored with less than the established minimum grade color of any variety; not over 15 percent all white berries and 5 percent berries of a green color; a count not over 175; not over 5 percent unsound berries; and with the name of the variety plainly stamped on the label. Altogether the sales company created about 45 brands, most but not all of them under the Eatmor name.

Independent growers and private distributors were using labels since before the turn of the

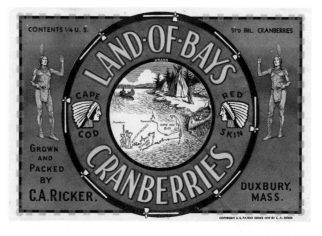

132

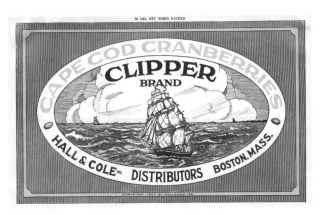

century, but they did not distinguish varieties to such a fine degree as did Eatmor. Usually, buyers of a particular crop knew what they were getting by the reputation of the packager. Some distributors, such as Beaton and Colley, did have several varieties each. But others, like Handy & Hennesey (Eagle Holt Brand), Ricker (Land of Bays), and Manomet Cranberry Co. (Big Injun), generally sold their cranberries under one label. Some labels represented distributors who were not growers, such as Austin, Nichols & Co. (Sunbeam).

Carver grower Eino Harju harvested, marketed and distributed his Pride of Carver Brand cranberries single-handedly: "In 1934 I bought floats and seconds that didn't make it through the separator. I cleaned them and sold them to New Jersey canners, to Ocean Spray, Stokely and Hills Bros. canners. I'd take a ride to Detroit, Albany, Montreal and Chicago in the summer and line up some sales."

By the 1950s, when processed cranberry products began consuming a larger share of the crop, the number of brand names and varieties decreased because the need to know a processed berry's life history was not so critical. And when consumer-size cardboard packages and cellophane bags became popular, the need to ship berries in crates to the retailer or distributor was eliminated, thus making labels obsolete. Berries could be packaged tightly in small quantities, kept fresh and be visually inspected by the consumer without being touched.

Today only two companies package and distribute fresh cranberries nationwide: Ocean Spray and Decas Cranberry Company, which distributes under the Paradise Meadow label. Both companies use polyethylene bags in 12- or 16-ounce packages. Ironically, putting up cranberries in one-pound consumer-size packages was tried 80 years ago by Wareham's Fuller-Hammond Company (bottom left), but it was 50 years too soon. (Labels courtesy of Wilho Harju, Larry Cole, Ocean Spray Cranberries, Inc. and Middleborough Public Library)

Ocean Spray's three-story Hanson plant in the 1920s. Only two decades earlier this had been a Hanson fire house. In 1910 Marcus L. Urann converted it into his Central Packing House, with the capacity of storing 50,000 barrels of fresh fruit at one time. By 1912 it was a fledgling canning operation called Ocean Spray. (Ocean Spray Cranberries, Inc.)

Ocean Spray's North Carver plant, formerly one of the New England Cranberry Sales Company's four screenhouses, ca. 1958. (Ocean Spray Cranberries, Inc.)

The Barnstable freezer, 1930s. The employees' rooming house ("The Fish Hook Club") is at the center, the freezer is at the far right and the executives' quarters and garage are at the far left. Berries not for immediate canning were sent to grower-owned freezing plants. It was said that freezing cranberries at their peak quality improves them by mellowing the flavor. Other freezers were at Chatham, Onset, Hanson and Harwich. (Ocean Spray Cranberries, Inc.)

The North Harwich plant in the 1920s. (Ocean Spray Cranberries, Inc.)

Three stages in the making of cranberry sauce. The bottom photograph was taken inside the Hanson plant shortly after it opened in 1912. The cranberry sauce was prepared manually and hand-filled and hand-sealed in #10 cans. The billing was done from an office Urann called a "board on a barrel." At right, a 1930s view of the berries traveling up to the top of a conveyor, where they were weighed and poured into the steel cooking kettle (background), which had a capacity of 500 pounds. Here the berries were cooked into a boiling sauce, which was then moved through stainless-steel pipes to the floor below, where the finisher removed skins and seeds. The sauce next traveled to the second-cook kettles (far right photograph, 1948) where it was mixed with sugar and jellied. (Ocean Spray Cranberries, Inc.)

136

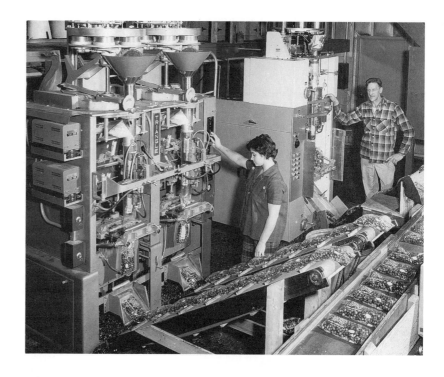

A polybag machine at Ocean Spray's Middleboro plant in the 1960s. The bags are used to package fresh cranberries. The machine forms, fills, cuts off and seals the bags. (Ocean Spray Cranberries, Inc.)

Top left: Labeling cans at the Onset plant in the 1940s. (Ocean Spray Cranberries, Inc.)

Filling one-gallon bottles of cocktail juice at the Hanson plant in the 1950s. To extract the juice, 350 tons of pressure is applied to 1500 pounds of frozen berries for 35 minutes. (Ocean Spray Cranberries, Inc.)

Part Three

Working Together

Overview

Christy Lowrance

In 1905 there were 4700 acres of cranberry bogs in Barnstable County and 6240 acres of bog in Plymouth County. Cranberries had become a million-dollar crop. This expansion created a need for information, so at the request of the growers' association, the state legislature established the University of Massachusetts Cranberry Experiment Station in East Wareham.

In the early 1920s, boxes replaced the big wooden barrels in which the cranberries were packed and shipped, and in the 1940s machine pickers replaced scoops for harvesting. Cellophane packages were introduced in the 1930s.

> Then they got expensive packaging machines. They said the transition would be three to five years, but that year we had to switch over. It all came quickly in one season, bingo! Once the village stores began to handle cranberries in packages, they found it was much easier than measuring them out. (Dud Eldredge)

Perhaps the most important addition to the industry in the twentieth century, however, was the Ocean Spray cooperative, created in 1930 by a merger of the A.D. Makepeace Company, New Jersey grower Elizabeth Lee's Cranberry Products Company and Marcus L. Urann's Cranberry Canners, Inc.

> At first, there was a lot of caution among the well-established men. They didn't join Ocean Spray right away because it couldn't offer the enticement, but those who were

The Ocean Spray cranberry stand. Built in 1933 on Routes 6 and 28 in Wareham, the 38-foot-high, 9-foot-diameter bottle drew 250,000 visitors in its first year, many of whom bought juice cocktail, sauce and cranberry frappes before continuing on their way. According to press releases about the bottle, many patrons "inquired where the products might be bought in their own home towns," which encouraged Ocean Spray to plan 100 more giant roadside bottles between Portland, Maine, and Philadelphia. None of these stands materialized, and the Wareham bottle, the flagship of the proposed fleet, ceased operating in 1943 and was torn down shortly afterwards. (Ocean Spray Cranberries, Inc.)

Opposite page: New Jersey NCA members, Walter Sloan, 80, from Chatsworth; Howard Buckalew, 81, from Lakehurst; and Andre Cottrell, 80, from Jackson Mills, in November 1947, examining the new cellophane packaging. Cellophane was part of a new and successful marketing strategy for cranberry growers. (Ocean Spray Cranberries, Inc.)

> taking over from their parents, or parents at the end of their careers, or young fellows starting out were interested. (Dud Eldredge)

The cooperative was able to give the same marketing advantages, as well as maximum return, to both large and small growers. It proved to be a providential move that enabled the industry to ride out the economic vacuum of the Great Depression. Harwich grower John Hall, whose great grandfather, Emulus Hall, was a grower in Henry Hall's era, recalls,

> ...the cranberry industry was pretty good during the Depression. It's about the only

A.D. Makepeace Company workers loading a shipment of berries at Tremont, 1920s. (A.D. Makepeace Co.)

United Cape Cod Cranberry Company, #1 Bog in Bryantville, 1943. Blondie Smith, in the foreground, gets boxes ready for harvesting. (Ocean Spray Cranberries, Inc.)

thing that was. A good cranberry bog was a paying proposition.

John Hall's father dealt scrap metal, farmed and grew cranberries:

> He always had bogs. We never went to school until after the harvest was over; it was how you got your clothes. When I was a kid everybody went and picked for other people. We picked with Smith and Hammond, Beaton, Urann—it was a family deal.

Hall bought his own bogs in Middleboro and Harwich about 30 years ago and was a charter member of the National Cranberry Growers' Association.

During the 1930s, there were about 800 growers in Massachusetts, and many of them belonged to cranberry clubs that met regularly for informal suppers and information sharing. Discussions back then often centered on new sprayer applicators, false blossom disease, bog tools and cooperative land purchases.

In a report given at one club meeting, Barnstable County extension agent Bertram Tomlinson discussed land and chemicals, issues that have continued to affect

the industry. He remarked that the loss of land—from 4331 acres in 1925 to 3533 in 1936—was a serious problem for the local industry, and he noted that,

> ...it seems the use of chemicals in combatting the various weeds offers the most promising solution of weed difficulties....Hand work in combatting weeds on a large scale is out of the question due to excessive cost.

World War II proved to be a considerable boon for cranberry growers. In one year the federal government purchased 40 percent of the national crop, which was canned, processed, dehydrated and served fresh to some 13 million servicemen and -women. Unfortunately, the civilian market was neglected during those years. The war's end sent prices down, forcing a number of growers out of business.

The family of Arthur Handy, a Cataumet grower and former president of the Cape Cod Cranberry Growers' Association, had been involved with cranberry culture for generations:

> My father wanted me to come home after World War II. He had 40 acres and now

we've built it to a little over 200 acres. My son Brian is the manager. There have been a lot of changes. When I came home in '46, there was little mechanization. People were still using a wheelbarrow to spread sand. We took old Model A Fords and cut everything off but the engine and the wheels, and carted sand onto the bogs with them. Then special machines came into use in the late 1940s.

Most growers in the forties had small parcels of land, and mechanization required a considerable amount of capital. Thus it was more practical for many to sell their land for its real estate value than risk farming at a loss.

Despite insect pests, the loss of acreage and the growth of the cranberry industry in other states (notably Wisconsin, which was harvesting more per acre using the new water-picking method), there was optimism that the industry in Massachusetts would grow. Hopes were especially high that Ocean Spray's aggressive marketing and new products would make the cranberry a year-round item rather than just a seasonal specialty.

Those hopes were dealt a major setback on November 9, 1959. On that day, Arthur Flemming, Secretary of Health, Education and Welfare, informed the nation that some Washington state and Oregon cranberries had been contaminated with the herbicide aminotriazole, which was being studied as a possible carcinogen. Although cranberry growers were the smallest users of the herbicide, and although the tainted berries were not from New England, the damage had been done. Approximately 79 percent of canned sales and 63 percent of fresh berry sales were destroyed on that day, still remembered by growers as "Black Monday."

Like a forest made stronger by a destructive fire, the cranberry industry was at first devastated but ultimately improved by the crisis. "In all catastrophic events, if you survive, you're stronger," remarked Ed Gelsthorpe of East Dennis, who became president of Ocean Spray shortly after the government's announcement. "If we hadn't had that we might still be muddling along." The industry was also helped by $8.5 million in reparations to growers ordered by President Eisenhower.

Ocean Spray's response to the aminotriazole crisis was the development of new products, particularly an improved cranberry juice cocktail:

People did not relate their fears to the cranberry juice cocktail. Cranberry sauce had always been flat in sales and seasonal. At that time, Ocean Spray was selling the cranberry juice cocktail in a container similar to

the old fashioned pop bottle. Sales were minuscule. We were looking for ways to broaden our base. The problem was the berries were tart. We took the problem to a consulting research firm in Boston, and did a fairly simple thing by making it more dilute, more sweet and adding new graphics. It took hold first on the East and then on the West Coast.

The cranberry scare of 1959 had another effect on the industry: It made it political. The growers' association hired an executive director, an advisor and a lobbyist. According to Sandwich grower Doug Beaton, the crisis,

> ...awakened us. It was the key to the planning we had to do, the commitment we had to make to ourselves, the environment and the industry. It focused us and gave us new direction...we're working with Mother Nature and the politicians.

This new political sophistication was well timed. It put the growers in a better position to weather the environmental storms that were to come in the seventies and eighties.

On November 19, 1959, photographer Ted Polumbaum was sent by Life magazine to cranberry country to document the effects of the "cancer scare" on the industry. Above, Francis "Brud" Phillips tells how he stands to lose $7500, his entire crop's earnings. At left, at the Ocean Spray plant in Onset, a man levels off a bin of 5000 barrels, or 50,000 pounds, of frozen cranberries that have no where to go. (Life magazine photographs by Ted Polumbaum)

143

The Cranberry Experiment Station

Robert Demanche

In 1892 Silas Besse planted his bog at Spectacle Pond in East Wareham and picked his first crop three years later. To control insects he set out 100 torches with 10-inch plates smeared with tar and molasses underneath each. Still, he lost almost a third of his first crop to fruit worm. In 1906 frost destroyed 90 percent of the crop.

Like Besse, many Cape and Plymouth County growers suffered the ravages of frost and insects. Frost had been

Above: Walter Bumpus, taking Dr. Henry J. Franklin on a tour in Mr. Makepeace's new automobile, ca. 1907, shortly after Franklin began his work at the Experiment Station. (Middleborough Public Library)

The Massachusetts Cranberry Experiment Station, as it looked in the 1930s. (Massachusetts Cranberry Experiment Station)

with them always and could be controlled by flooding, but by the late 1800s, insects, weeds and funguses seemed to have stepped up their attacks on the cranberry. Perhaps it was the close concentration of vines on the heavily cultivated bog that fostered an increase in pests, but whatever the reason, the growers needed help. Their determination to find it led in 1910 to the establishment of the Cranberry Experiment Station.

The experiment station is a department of the College of Food and Natural Resources of the University of Massachusetts. It conducts research and provides information to growers on cranberry cultivation, management and harvesting—research the typical grower hasn't the time, the money or the expertise to carry out. An act of the Massachusetts legislature brought the station into existence. A pledge by Cape Cod Cranberry Growers' Association members of one cent per barrel shipped in 1909, totaling $834, helped pay the salary of its first director, Dr. Henry J. Franklin.

To house the station, the state bought Silas Besse's 12-acre bog, now known as the State Bog, and constructed a research building in which Dr. Franklin lived until his marriage seven years later. He was the only worker at the station until 1913, when his longtime assistant, Joe Kelley, signed on.

The early history of the experiment station is in large part the history of Dr. Franklin, whose work on insects and frost would make him the foremost cranberry expert of his time. Although trained as an entomologist, Franklin attacked the subject of cranberries on many fronts. By 1914 he had set out 53 experimental plots on the State Bog: 24 devoted to fertilizer studies, 16 to fungus diseases, 6 to insects, 5 to sanding and 2 to harvesting.

On the basis of his observations of insects, Franklin was able to map out the life cycles of most of the cranberry's predators and determine for each the point at which populations became large enough to warrant action against them. He established three principles which still govern insect control today: First, monitor the bog regularly; second, identify and count what you find; and third, act only when the "action threshold" is reached.

Up until a few years ago, spring frosts were responsible for more crop loss than were insects. The problem for growers, once the winter flood was off the bog, was predicting accurately when a frost would occur and getting the flood going before the temperature started to drop. Franklin took much of the guesswork out of frost predict-

ing. Beginning in 1912, he and some helpers set up stations in Marstons Mills, Norton, Halifax, South Hanson, Pembroke, South Carver and at the experiment station to gather data on temperature, dew point, wind speed, cloud cover and precipitation. From this information he developed prediction formulas that were the basis of the experiment station's telephone frost warning system, instituted in 1920. This system, refined by Franklin through the years, remains basically unchanged today.

In 1915 Dr. Neil Stevens, a plant pathologist with the U.S. Department of Agriculture, joined Franklin and Kelley on a seasonal basis to study false blossom disease, a major problem for the growers, and how temperature affects the cranberry's keeping quality. Later Dr. H.F. Bergman, another seasonal worker from the USDA, studied how the oxygen content of winter flood waters

George Rounsville, the Experiment Station's bog foreman, and an unidentified helper, hand-planting vines on Section #3 at the State Bog, 1940s. (Massachusetts Cranberry Experiment Station)

affects the vines. In 1946 Dr. Frederick Chandler joined the station to work on fertilizers, drainage and new cranberry varieties.

Disseminating the work of Franklin and his colleagues was the mission of the county extension services. The Upper and Lower Cape Cranberry Clubs, started in 1935 by Barnstable County extension agent Bert Tomlinson, were one attempt to fulfill this mission by providing an opportunity for growers to learn about and discuss new cultivation techniques, products and machinery. Plymouth County soon had its own cranberry clubs.

Another collaborative effort by the experiment station, the extension service and the growers was the G.I. Cranberry School, a two-year program of lectures and discussions begun in 1946 to teach returning veterans the fundamentals of cranberry growing. Held at Ellis D. Atwood's screenhouse, the course averaged 136 students per session, much more than extension specialist Richard Beattie had anticipated. Similar classes at Barnstable High School also attracted many would-be growers.

Dr. Franklin retired as station director in 1952. His successor was Dr. Chester Cross, a botanist who had come to the station in the late 1930s. Cross developed salt, Stoddard solvent, paradichlorobenzene and the arsenicals for use as herbicides. He also perfected the use of kerosene as a weed killer—first discovered by station researcher Dr. William H. Sawyer in the mid-1930s.

In a 1977 interview in the UMass alumni magazine, Dr. Cross stated his belief that the government banned some pesticides without sufficient evidence of their long-term harm and without considering their benefits. He said that agricultural chemicals, both pesticides and fertilizers, are at the root of the station's research. "Controlling pests is not the problem; we know how to do that. The problem is finding chemicals that do their jobs and then disappear in the aquatic environment."

Dr. Franklin's retirement created the need for an entomologist at the experiment station. This was Prof. William Tomlinson, who, in addition to his work on insect pests, conducted studies on using honeybees in cranberry pollination. Dr. Bert Zuckerman, a plant pathologist hired in 1955, studied the nematode, a microscopic subterranean worm that can kill cranberry insect predators. He also pioneered aerial application of fungicides.

Other researchers who began under Cross were Dr. Wes Miller, who studied agricultural chemical residues; Dr. Charles Brodel, who worked on insect control strategies; and two researchers who still work at the station: Dr. Karl Deubert, who investigates the breakdown and movement of pesticides away from the site of application, and Dr. Robert Devlin, who studies plant growth regulators and experimental herbicides.

To keep up with the increasing use of heavy construction equipment and mechanical harvesters on the bogs after World War II, the station hired Stan Norton, an agricultural engineer, in 1957. Norton designed and perfected the low-gallonage sprinkler system as well as several types of bog machinery.

In the 1920s, a Washington cranberry researcher discovered that water continuously sprinkled onto the vines during cold weather radiates heat as it freezes, keeping the plant warmer than the critical low temperature that would destroy it. At first growers installed sprinklers on only a very few acres of bog. A system of that era might be constructed of 2-inch diameter aluminum pipe drilled with small holes, propped a few feet above the bog on two-by-fours. Norton's low-gallonage system, made with 3/4-inch diameter flexible polyethylene tubing, was cheaper and easier to use and required much less water to achieve its goal.

On the night of Memorial Day, 1961, a severe frost wiped out one-third of the Massachusetts cranberry crop.

Dr. Franklin, Dr. Sawyer and Dr. Bergman, probably working on insect slides in their one-room office at the station. This room, known as the packing room, also housed the separating machine and blower. Because a fire destroyed Franklin's first office and all his research papers, he kept a post office safe in the corner of the room and locked up his data every night. (Massachusetts Cranberry Experiment Station)

This disaster was enough to convert many growers to the low-gallonage sprinkler system, and by 1967 more than half of the total bog acreage was "under sprinklers."

Wisconsin growers had harvested "on the flood" since the turn of the century. In fact, water picking had helped to make Wisconsin the second largest producer of cranberries in the nation. Massachusetts growers were slow to adopt the new method, however, and it was only in the early 1960s that David Mann of Buzzards Bay became the first to water-harvest his crop commercially.

Experiment station researchers perfected the walking water wheel, and they demonstrated that, though water harvesting would increase yields by at least 20 percent, berries picked this way had a lesser keeping quality than dry-picked ones and so could not be sold as fresh fruit.

In 1981 Dr. Cross retired and Irving "Dee" Demoranville became the experiment station's new director. Demoranville began at the station in 1952, working on weed control and cranberry varieties. He later succeeded Beattie as cranberry extension specialist. Demoranville's tenure has seen some interesting changes, including the introduction of Integrated Pest Management (IPM) and a new focus on research in biological control—controlling pests by using their natural enemies.

During Franklin's early years at the station, only a few common chemicals, such as lime and copper sulfate, were available to fight insects, weeds and other pests.

Flooding to hold down the number of insects and good drainage to discourage weed growth were the main control methods. By the late 1930s, kerosene was found to kill weeds without hurting the dormant cranberry plant, and later station researchers demonstrated the effectiveness of many other chemical pesticides.

By the 1970s, people throughout America began to question what effect agricultural chemicals had on air and water quality near the growing areas. These concerns led to the development in the late 1970s of Integrated Pest Management (IPM), which federal and state governments have actively promoted throughout the 1980s. Cranberry station researcher Sherri Roberts developed the station's IPM program in 1983. According to Roberts,

> IPM means that you grow your crops as economically as possible, using whatever means you have at hand, whether cultural, biological or chemical, but that you use chemicals only when necessary.

Under IPM the action thresholds Dr. Franklin developed are more precisely defined for both the economics of insect damage and the precision of "when to spray."

> For example, the threshold for weevils is 4.5 per 25 sweeps of the net. If you sweep your bog with your net and find 4.5 weevils, you spray. If you find only 4.2, you keep an eye on things, but you don't spray. (Sherri Roberts)

Now, about 25 percent of Massachusetts' cranberry acreage is under IPM, a figure that is increasing. Most of this acreage is handled through IPM programs run by companies such as Roberts' Cranberry Consultants, Ocean Spray and Decas Company. Station entomologist Anne Averill points out that this was the University of Massachusetts' original intention:

> First, establish the program at the station and then teach others so that they would spin off to form their own companies. The station would then devote more time to research.

Two areas of biological control look promising in the near future. The experiment station, in collaboration with Ocean Spray and other companies, is studying the

use of B.t. (*Bacillus thuringiensis*), a naturally occurring bacterium, against leaf-eating caterpillars. And Ocean Spray's recently opened research lab is reviving research into the use of nematodes to control harmful soil insects. Dr. Frank Caruso, who joined the station in 1985 as plant pathologist, is looking for chemical, cultural and biological ways to control a widespread, more recent problem, the "root rot" fungus.

The tripling of cranberry production since 1960 has revived research on fertilizers, which Franklin had originally dismissed as unnecessary. He was right at the time; when he carried out his studies, growers produced fewer barrels per acre and the amount of nutrients in the soil exceeded what the plants needed. Now, however, since more nutrients are going to support higher yields, leading to rapid depletion, scientists are giving fertilizers a second look. For example, Carolyn Demoranville, a Ph.D. student at the station, is investigating the benefits of fertilizer made from fish parts discarded after filleting.

In 1910, when the Cranberry Experiment Station opened, the average yield was about twenty 100-pound barrels of cranberries from each acre of bog. Today the average yield is over 150 barrels per acre. As the primary research organization for cranberry cultivation in Massachusetts, the experiment station has played a significant role in helping the grower achieve this high yield.

Franklin's work on frost and insects was the main reason production doubled from 20 to 40 barrels per acre during his tenure. Since then water-harvesting, Norton's work on the low-gallonage sprinkler system, and new methods of pest control have brought yields to their current levels.

But increased production is not the only thing the station is there for. It has disseminated the latest information on cultivation and has acted as a liaison between growers, industry and government.

Growers are farmers and businessmen, not scientists, and no matter how successful, few of them have the resources to do what the experiment station can do. As Demoranville points out,

> Is a fellow who's got a nice bog and who's taking care of it going to go out there and experiment with some kind of machine that he doesn't know what it's going to do, that may roll his vines up like a carpet?...Not at $50 a barrel and 150 barrels an acre, he's not.

The relationship between the growers and the experiment station is a cooperative one, resting on the positive tension between new ideas and methods proven and established long ago. One long-time grower remarked, as a bulldozer dug into his bog,

> Last year the people at the cranberry station said I was getting strawberry weevil on this bog, and if it took hold it would be very expensive to get rid of. I'd probably have to tear off the top layer of the bog and replant it. I didn't believe them. Then I started to lose my vines. This year I believed them.

Anne Averill puts the station's mission into focus:

> I have a great respect for the growers. They have great instincts and they have generations of knowledge behind them. What I can bring is my 15 years of experience as an entomologist and my knowledge of other fields of agriculture. They know their bogs inside and out. I can share the knowledge I get from visiting 500 acres of bog a week. All that I can offer them is another set of eyes, another viewpoint on which they can base their decisions.

Irving "Dee" Demoranville, current director of the experiment station. A Freetown cranberry grower as well as an horticulturalist, Dee started at the station in 1950 as a student in the research lab. Since becoming director, he has led the station in its efforts to name and destroy pests. Demoranville feels his highest achievements are developing three new hybrid cranberry varieties (the Franklin, the Bergman and the Pilgrim) and making herbicides, such as Casaron, available to growers. (John Robson photograph)

Drs. Chester and Shirley Cross. Chet Cross brought both activism and diplomacy to the station during his 20-year tenure as director (1952–81). His specialty was the study and control of the grasses that infest bogs. An outspoken critic of the government's handling of the aminotriazole crisis of 1959, Cross also talked to groups in the region on agriculture and the cranberry industry. In the late '60s he worked with state officials to build the station's extensive laboratory, which helped the station attract researchers from Poland, Japan, the Philippines and elsewhere. He died in 1988.

Shirley Cross worked closely with her husband and helped manage their Sandwich bog. A biologist by profession, she has also illustrated many of her husband's manuscripts and serves as president of the Thornton Burgess Society. (Ocean Spray Cranberries, Inc.)

Cooperation and Competition: The Ocean Spray Story

Dan Georgianna

As long as people have farmed the land, they have cooperated with each other. Farming can be hard, lonely work; many hands working together make planting, harvesting, maintenance and construction easier. Farming is also a risky occupation; the burdens of drought, disease and pestilence are lighter when shared. Finally, time is the enemy of the farmer and a powerful ally of the buyer in a market glutted with a perishable harvest; joining together gives farmers bargaining strength when faced with high prices for supplies and low prices for crops.

Cooperatives fit easily with the politics and ideas of nineteenth-century America. The great advances in cultivation and harvest yielded bumper crops that depressed prices and ruined hard-working farmers. Dependent upon business but resentful and suspicious of its power, farmers saw businessmen as exploiting them in the same way industrial capitalists exploited factory workers. The robber barons, giving little yet receiving much, contradicted the farmers' belief in equality and democracy.

Thus from their beginnings in the nineteenth century, cooperatives were viewed by farmers as good pitted against evil. Only by joining together could farmers secure relief from businessmen, who prospered as they suffered.

By the end of the 1920s, almost all American farmers belonged to some form of co-op, through which they borrowed money, bought supplies and marketed their crops. Since then, however, the number of co-ops has decreased at roughly the same rate as has the number of farms, even though their market power has increased. Over the past 30 years, farm co-ops have doubled their share of the market for agricultural produce.

PICKERS   WEST WAREHAM, MA

Cooperative Marketing of Cranberries

By the latter third of the 1800s, cranberries, though generally sold only from Thanksgiving to Christmas, had become part of the national fruit market, handled through brokers, commission houses and wholesalers. The berries were sold on consignment, a method fraught with danger for the farmer with a perishable commodity selling in a distant market to unknown buyers. Too often, the farmer received only a little money, perhaps accompanied by a note filled with regret that the market didn't pay more.

Occasionally a grower could get satisfaction. A story still current among the growers, although it happened almost a century ago, is that of Ben Shaw, a cranberry grower in Rocky Meadows around the turn of the century. According to Frank Cole, who sold him crates, Shaw shipped some berries on consignment to a broker in New York, who sent him a check for much less than Shaw expected, along with a note explaining that the market was saturated. Shaw decided to see for himself if this was true and rode the train to New York. He presented himself as a buyer to his broker, whom he had never met, and made a low offer for a few barrels of his own berries. The broker refused the offer saying that the market was firm with rising prices. Shaw then reached into his pocket as if to pay for his purchase, but instead slowly drew out the check he had received and the broker's note on the sorry state of the market.

A more practical way for the growers to receive full value for their fruit was to share information and coordinate sales in the wholesale market. In 1888 A.D.

Makepeace and other growers on Cape Cod voted to form "an organization to be known as the Cape Cod Cranberry Growers' Association, whose object shall be to promote the interests of the growers of cranberries in Plymouth and Barnstable Counties." While not a commission house, a brokerage or a marketing cooperative, the growers' association nevertheless attempted to standardize, promote and place the product.

A more direct effort was the formation of commission companies jointly owned by growers. These didn't promote a uniform brand name, nor did they pool the crop; instead, they paid growers the value their berries fetched on the market less a 5- to 7-percent commission. The most successful commission company was the Growers' Cranberry Company, formed by growers in New Jersey and Massachusetts in 1895. It sold about one-quarter of the crop in these growing regions, mostly in the major cities of the East Coast.

In 1902 cooperative marketing of cranberries took a great leap forward with the arrival of A.U. Chaney, operator of a modest wholesale fruit business in Des Moines, Iowa. Chaney saw the farmers' problems with unscrupulous agents as a business opportunity and with two other buyers bought the entire Wisconsin crop in 1905. The next year saw a bumper crop, which sharply reduced prices. Impressed by Chaney's success, Judge Gaynor, an influential Wisconsin grower, asked him to draw up a plan for a marketing cooperative. Fear of a market glut quickly sold other growers on the idea, and in 1906 Chaney and Gaynor formed the Wisconsin Cranberry Sales Company.

151

PAUL REVERE BRAND
ONE FOURTH U.S.
STANDARD BARREL
TRADE MARK
CAPE COD
CRANBERRIES
PACKED FOR THE
AMERICAN CRANBERRY EXCHANGE
NEW YORK | PACKERS NO. | CHICAGO

Paul Revere Brand, a pre-Eatmor label from the American Cranberry Exchange, ca. 1920. Paul Revere berries were Early Reds, at least 85 percent colored, with a count not over 110 per cup, and fit for 10 days' travel. (Massachusetts Cranberry Experiment Station)

Beaver Brand label, ca. 1910. The Wisconsin Cranberry Sales Company, formed in 1905 by A.U. Chaney and Judge Gaynor, was the first cranberry marketing cooperative. In 1907 it joined with the New England Cranberry Sales Company and the New Jersey Cranberry Sales Company to form the American Cranberry Exchange. (Courtesy of Wilho Harju)

Arthur U. Chaney, 1874–1941, organizer and president of the American Cranberry Exchange. At his death, the directors of the Exchange wrote of him:

In recent months he had labored under increasing nervous tension, but his thirst for work and responsibility did not lag. The activities of his last few days were characteristic. Satuday afternoon was spent in the office analyzing figures and tearing them to pieces. Long after the lights flickered in office windows he asked for a conference on market conditions in every corner of the country. Upon arrival home Saturday evening he slipped into an easy chair and with a sigh remarked, "The crop is marketed." Sunday he was at home, addressing Christmas cards, writing a friendly letter, and contemplating a well-earned vacation. During the night he lapsed into a coma. Early Tuesday evening he died . . .

Chaney was succeeded by his brother Chester as president of the Exchange, but no one ever succeeded his capabilities. "He earned his preeminence by long and intensive study of our problems, by good judgment, earnestness and square dealing," wrote the directors. ". . . He held the confidence of producers and distributors." (Courtesy of Clark Griffith)

Almost all of the Wisconsin growers joined in the first year. The company, a grower-owned marketing cooperative with Chaney as exclusive sales agent, promised an orderly market, paying each grower the same price for berries. No longer could buyers pit grower against grower.

In 1907 Chaney and Gaynor came east preaching the gospel of cooperative marketing to the growers of Massachusetts and New Jersey. As a result of this visit, the growers there formed the New England Cranberry Sales Company (located in Middleboro) and the New Jersey Cranberry Sales Company, both based on the Wisconsin model. These two co-ops then merged with the Wisconsin co-op to become the National Fruit Exchange, with Chaney as general manager. In a single year, Chaney had collected a majority of the crop into a farmers' marketing co-op with sales companies in each of the three major growing states.

For a few years, price-cutting competition between the National Fruit Exchange and the Growers' Cranberry Company, though harmful to the growers, was not ruinous, but then a bumper crop in 1910 caused an all-out

BEAVER BRAND
EXTRA FANCY
HIGHLY CULTIVATED WISCONSIN
CRANBERRIES
LATE KEEPING VARIETY
GROWN AND PACKED ESPECIALLY FOR FANCY TRADE
HAND ASSORTED AND UNIFORMLY PACKED
WISCONSIN CRANBERRY SALES CO.
GRAND RAPIDS, WIS.
PACKERS NO.
A.U. CHANEY CO
SOLE SELLING AGENTS
DES MOINES, IA.

price war between the two. The battle was short: Since the Exchange had most of the product, the confidence of the growers and A.U. Chaney, the Growers' Cranberry Company soon conceded and merged with the New Jersey Sales Company branch of the Exchange.

The Exchange was a well-organized marketing co-op owned and directed by the growers. Its organization reflected the farmers' democratic principles and their mistrust of business organizations. Each sales company had its own officers and board of directors. The directors of the Exchange were representatives of their sales companies, and the bylaws insured that no state would have a majority on the board. All elections followed the rule of one person–one vote. Membership fees were modest, and there was no capital stock. Each sales company was supported by a revolving fund of loans from the members, payable at a fixed interest rate with a fixed due date.

Members contracted to deliver their entire crop to the sales companies, and the Exchange handled marketing, grading, transportation, advertising and sales. The berries were collected into yearly pools according to grade, type and state and sold under the "Eatmor" brand name. Upon delivery growers received a small down payment, and when the crop was sold out they divided the net proceeds from the pool according to the quantity and quality of berries they delivered.

The Exchange's system solved many problems for growers. It regulated timing and place of sale, smoothing out the wild swings in price due to seasonal demand and perishability of the product. The pooling arrangement stopped growers from rushing into a profitable market and sharply reducing the price. Most important, the uncertainty of sales was reduced. Gone were the days of shipping on consignment to unknown, possibly dishonest dealers and brokers. Members could concentrate on what they knew best: growing cranberries. By regulating the supply to wholesalers, the Exchange succeeded in eliminating large price drops caused by glutted markets.

A.U. Chaney was smart, and he knew the cranberry business. Having been a successful wholesaler, he could predict the quantity and pattern of demand and sell the berries at the highest average price. At the beginning of the sales season, he would set an opening price that was low enough to start the berries moving and high enough to ensure that sufficient supply would be available at peak periods of demand. An accurate opening price, combined with careful placement of the berries at different times in various markets, would lead to a gradually rising price with all berries sold by year's end.

Success has a way of leading to failure in the business world. In the case of the Exchange, successful marketing brought larger profits for the growers but also encouraged larger crops. The excess supply in turn caused prices to drop. Ten years after the founding of the Exchange, the yearly supply of cranberries had almost doubled. Limiting supply wasn't possible. Then, as now, it was against a farmer's nature to reduce his crop. Furthermore, not all growers belonged to the co-op; any reduction in the Exchange's supply in order to raise prices would only encourage the independent growers to supply more.

The major problem was the limited demand for cranberries, which the public saw as a specialty item reserved for certain holidays. Advertising to encourage demand was an obvious solution, but the growers resisted. Without consumer loyalty, they knew that any increased sales from advertising could easily be seized by the independents, who wouldn't be paying any of the advertising costs. Generic advertising would reward the free riders.

The Exchange decided on a two-point strategy: Promote the Eatmor label, and increase its membership so that the costs of advertising could be shared by more of the growers. The gain to free riders would be galling but limited. Besides, the sharp increases in the harvest left little choice but to advertise even if the Exchange couldn't capture all gains.

In the fall of 1918, worried about another large crop and the shortage of sugar caused by the war, the Exchange spent $54,000 on a modest newspaper and magazine advertising campaign. At the same time, it encouraged dealers to display the fruit and hold the line against price cutting. As a result of these strategies, the retail surplus disappeared after Thanksgiving and by Christmas demand was five times greater than it had been the year before. Chaney was able to report that the $54,000 ad

Pheasant Brand, ca. 1930. These berries are the Centennial variety, of uniform dark color, a count not over 75 per cup, and fit for 7 days' travel. The Centennial was developed in 1876 in Holliston by George Batchelder. (Massachusetts Cranberry Experiment Station)

Chanticleer Brand, ca. 1915, a pre-Eatmor label of the New England Cranberry Sales Company. Chanticleers were Early Blacks, at least 90 percent colored, with a count not over 125 to 150 per cup, and fit for 15 days' travel. (Massachusetts Cranberry Experiment Station)

Below: A gathering of the first group of New England Cranberry Sales Company members outside their Middleboro office, ca. 1910. (Ocean Spray Cranberries, Inc.)

Marcus L. Urann, 1873–1963. M.L.Urann formed the United Cape Cod Cranberry Company in 1906 and by 1909 he had $500,000 worth of common and preferred stock and was on his way to becoming one of the largest growers in the state. In 1912 he began making canned cranberry sauce at his Hanson packing house, adopting the slogan "ready to serve" and using the Ocean Spray Brand label. Canned cranberry sauce was one of the first convenience foods, and by the time of Urann's retirement in 1953, 60 percent of the nation's crop went into its manufacture.

Until 1930, when the Cranberry Canners, Inc. cooperative was formed and he became its president, Urann owned his own canning plants and Ocean Spray was his private brand. During the war, the cooperative under Urann was awarded the government's agricultural "A" award for supplying the military with dehydrated cranberries. In 1946, the consulting firm of Booz, Allen, Hamilton analyzed the cranberry industry, finding that Urann was "right too soon…[and] had to wait for time to catch up with his view." They also noted that while "his intelligence and business capacity is almost universally admitted by both his friends and his foes … Nevertheless, it is an outstanding fact, as revealed by survey interviews, that more than two-thirds of the growers do not fully trust Mr. Urann or his motives."

Still, most growers agree that it was better to have Urann on their side. He convinced growers that outside canneries were their natural adversary. "If canners who are not growers get cranberries," he wrote in 1941, "the canning profit will enable them to depress the price of cranberries, and sell the canned goods at a profit to themselves. If growers do the canning, they get this extra profit and also protect the fresh cranberry market." Urann succeeded in protecting the market for the cooperative. Between 1938 and 1953 (Urann's tenure), the co-op's assets grew from $2 million to $12.5 million. (Ocean Spray Cranberries, Inc.)

Right: Early Ocean Spray can label, ca. 1930. (Ocean Spray Cranberries, Inc.)

vertising investment had returned $1 million in sales. He and the growers had solved the problem of fresh fruit marketing: advertising, judicious timing and support and encouragement for the retailers.

The Founding of Ocean Spray

Since almost all cranberry sales in the early 1900s were in the fresh fruit market, product quality was a major concern of the Exchange as well as a source of pride to the growers. The returns from advertising and consumer preference for the Eatmor brand were too valuable to lose. Exchange inspectors regularly checked berries for quality and settled disputes between growers and buyers.

There were two nagging problems with fresh fruit sales. One was the loss from dumping the lower-quality berries, which couldn't be sold in the fresh fruit market without harming Eatmor's image or creating competition for the quality berries. The other was the major change taking place in America's kitchens. Food preparation was shifting from the kitchen to the cannery, creating a growing market for cheaper processed berries.

Once again a problem became a business opportunity

for an aspiring entrepreneur. Marcus L. Urann, a creative grower and businessman saw the future of cranberries in cans and bottles.

Urann was born in Sullivan, Maine in 1873. In 1897 he graduated from the University of Maine, where he had been captain of the football team and founder of the national honor fraternity Phi Kappa Phi. He then studied law at Boston University and opened a practice in Easton, Massachusetts in 1900. From one of his first clients, he learned that cranberries were profitable and that good land was still cheap. In 1906, while Chaney was organizing the first marketing cooperative in Wisconsin, Urann bought his first bog.

Urann was easily converted to Chaney's gospel of a national cooperative and was a founder and member of the board of directors of both the New England Cranberry Sales Company and the American Cranberry Exchange. Although powerful men used to getting their own way, he and Chaney worked well together; both saw the necessity of cooperation as a defense against the ruinous competition of the market.

Yet they were not always of the same mind. In 1912, after failing to convince Chaney and the Exchange of the benefits of processing, Urann began canning his own berries under the brand name "Ocean Spray" and slowly built up a substantial business in canned sauce. His primary competition was John C. Makepeace, another major partner in the Exchange, who was canning berries under the "Makepeace" brand, and Elizabeth Lee of New Jersey, who owned Cranberry Products Company.

Since growers considered cranberries that were unfit for the fresh fruit market a near total loss, they sold them for practically nothing to the canners, who processed and sold them at low prices and still made a profit. This was a problem for canners like Urann and Makepeace because they were also major growers who wanted good prices for their berries as well as profits from canning. In a competitive market, they had to sell at low consumer prices even if they processed high-quality berries.

Urann's solution was a merger of the three major canners as a way to better control the price of berries for processing, and in the spring of 1930 he negotiated a merger with his rivals. Following his usual procedure of acting first and asking questions later, only afterwards did he inquire if such a merger would be legal. The answer, according to John Quarles, a young lawyer trying to make a living during the Great Depression, was a reluctant no—antitrust laws prohibited mergers in which the new company would control over 90 percent of the market. Undeterred, Urann told Quarles to find a way to make the merger legal.

Upon further investigation, Quarles came across the Capper-Volstead Act exempting agricultural marketing cooperatives from the monopoly restrictions of antitrust laws. If the canners formed a marketing co-op, their merger would be legal. The only problem was that the act was intended to help growers in the marketplace, not to reduce competition among processors, but as Quarles saw it, Urann and his partners were primarily growers and as such were entitled to the exemption. When the Justice and Agriculture departments gave their blessing to this interpretation, Quarles was sure that the courts would go along with it as well.

Quarles worked quickly to transform the three rival canners into a farmers' marketing cooperative. Urann, Makepeace and Lee agreed to turn over all their canning assets to the new cooperative while retaining control of their growing operations. This meant that Urann, who was contributing the majority of the assets, would have effective control of the new co-op as president and major stockholder. However, while Urann dominated processing, Makepeace grew more cranberries than anyone else in the world, and his family had long been the major

Two views of the United Cape Cod Cranberry Company plant in Hanson, ca. 1920. At left trucks wait at the docks for a load of Ocean Spray sauce; inside the plant, at right, cranberries are stewed in 500-gallon aluminum kettles. (Ocean Spray Cranberries, Inc.)

Top left: Bog-Sweets, the trademark carried by Elizabeth Lee's Cranberry Products Company. (Ocean Spray Cranberries, Inc.)

John C. Makepeace. At A.D.'s death in 1913 John C. took over the business, following the independent, innovative course set by his father. He established a box and scoop mill at Tihonet in 1925 and a canning plant in 1928. In the 1930s company men Everett Niemi and Ernie Howes developed a volume-controlled packing machine, which led to the company being the first to use cellophane packaging in 1942.

In 1930 John C., Marcus Urann and Elizabeth Lee formed the Ocean Spray cooperative. The next year the Makepeace "poison plant" opened for the manufacture of pyrethrum and other pesticides. In the 1940s the company was one of the first to adopt aerial spraying, using its own Stearman biplane.

John C. continued the work he and his father had started on dehydration, and in 1944 the entire crop was dried and shipped to the military overseas. For this effort the company won a USDA Achievement A award in 1945. John C. Makepeace retired in 1958.

Right: Makepeace Cranberry Sauce, 1929. (Ocean Spray Cranberries, Inc.)

Opposite page, left: The interior of the A.D. Makepeace Company canning plant, 1928. (Ocean Spray Cranberries, Inc.)

Opposite page, top right: Makepeace Crannies, from the 1940s. "Crannies make ten times as much sauce as an equal weight of fresh Cranberries," according to the package. "Always boil HARD. Do not stew or simmer." (Ocean Spray Cranberries, Inc.)

Opposite page, bottom: The Plymouth processing plant for Ocean Spray cranberry sauce, ca. 1940. (Ocean Spray Cranberries, Inc.)

force in the industry. Like Urann, Makepeace was a powerful man with tremendous energy and a tough businessman. He didn't find it easy to give Urann control over any part of his operation. Further, Makepeace couldn't be sure that Urann wouldn't continue to use the co-op to his own advantage once his opposition had been eliminated.

According to Quarles, the atmosphere when they all met to sign the papers was not one "...in which much would be taken on faith." Makepeace asked Urann if he had changed the name of his growing operation so that the new co-op could use the name Ocean Spray on its labels. Urann answered no, but said that it would be done.

Makepeace asked, "When?"

Urann replied, "Sometime soon."

Makepeace, sensing a double cross, started to walk out, declaring that the deal was off and demanding his papers back. Quarles replied that he couldn't give him the papers that the others had signed and that he would either destroy all the contracts or hold them in the hope of future resolution. Makepeace reluctantly agreed to allow Quarles to mediate, and all parties retired to separate rooms. Quarles acted as go-between in the effort to bring the two strong-willed men back together.

Finally Urann signed over the brand name to the new

co-op, and on Thursday, August 14, 1930, Cranberry Canners, Inc. came into existence. Urann, as president, controlled 50 percent of the new company, Makepeace, as treasurer, controlled 25 percent, and Lee and a few other growers controlled what remained. The common stock distribution was arranged so that Urann, Makepeace and Lee each had veto power. Although the name of the co-op was to change to the National Cranberry Association in 1946 and then to Ocean Spray in 1959, its products were labeled "Ocean Spray" from the beginning.

Almost immediately there were problems with the new co-op. A sharp rise in prices for canned berries led some buyers to complain to the courts about the new company's market power. Quarles exorcised that problem by producing the blessing he had received from Washington. A more significant omen was the relationship of the new co-op to other growers, most of whom belonged to the Exchange. Urann couldn't demand all of the new members' crops since he didn't have an outlet for fresh product and didn't want to compete with the Exchange. So he decided to settle for 10 percent, roughly the estimate of the share that wasn't suitable for sale as fresh fruit. Even under this arrangement, independent growers (those not in the Exchange) couldn't supply enough fruit to satisfy Urann's plans for growth. What he needed was a way to lure some of the growers away from the Exchange and into Ocean Spray.

The solution was for the sales companies of the Exchange to join the new co-op. In 1934 the New England Cranberry Sales Company joined Ocean Spray, contracting to supply 10 percent of its pool to Ocean Spray for processing. The New Jersey and Wisconsin sales companies followed in 1938 and 1940. Growers in the Ex-

MAKEPEACE
CRANNIES
REG. U. S. PAT. OFF.

SLICED • • •
**WHOLE, DEHYDRATED
CAPE COD CRANBERRIES**

Crannies are whole, so can be used for the same purposes
as fresh Cranberries, retaining the natural flavor and color.

PRODUCT OF
A. D. MAKEPEACE COMPANY
WAREHAM, MASSACHUSETTS
U. S. A.

NET WEIGHT: ONE POUND

PRINTED IN U. S. A.

change could deliver berries to Ocean Spray, but the Exchange would be the seller, preserving the growers' contracts with the Exchange to sell the entire crop. As an added bonus, Urann and Makepeace, still directors of the Exchange, did not have to pay overhead to the Exchange for selling their own berries to Ocean Spray.

To many of the growers, the arrangement between the Exchange and Ocean Spray was an unholy alliance. Though in different markets, canned product competed with fresh berries. Growers who had profited from the orderly fresh fruit marketing strategies of Chaney were reluctant to risk this success for the small gain to be had from selling a few damaged berries. They stood to lose too much from a collapse in the fresh fruit market, and besides, most of them didn't trust Urann.

The Battle of the Co-ops

Unlike other farmers, cranberry growers didn't suffer from the Great Depression of the 1930s; prices generally stayed high and the berries sold out every year except 1937, when a large crop coincided with the Depression's worst year. That year the Exchange couldn't sell all of its berries on the fresh fruit market and was forced to deliver about half of its crop to Ocean Spray for freezing and future processing. As a result, Exchange members weren't

On the boardwalk in Atlantic City, N.J., 1936. W.M. Campbell (wearing hat) hears from M.L. Urann how Ocean Spray brand cranberry sauce has just gone over the top. (Ocean Spray Cranberries, Inc.)

Opposite page, right: The Cranberry Army Pool, ca. 1941. At the beginning of World War II, Ocean Spray organized the army pool to provide dehydrated cranberries for the armed forces. The first growers to join were Mrs. Lincoln and Mrs. Clapp of Scituate. (Ocean Spray Cranberries, Inc.)

Opposite page, left: John C. Makepeace (second from right), Marcus Urann (center) and other NCA executives, admiring the company's new shipping containers, 1940s. (Ocean Spray Cranberries, Inc.)

Opposite page, bottom: Evaporated cranberries, 1940s. Ten pounds of fresh berries were dehydrated into a 4- x 4- x 3-inch brick weighing one pound. This brick made approximately 25 pounds of cranberry sauce, or enough for about 100 servings. (Ocean Spray Cranberries, Inc.)

paid for their berries until Ocean Spray sold them all, which augured ill for the Exchange and boded well for Urann and his belief in processing.

Another blow to the Exchange was the death in 1941 of A.U. Chaney and the appointment of his brother Chester as General Manager. Though he had been assistant manager since the beginning of the Exchange, Chester didn't have his brother's genius, nor did the growers have the same level of confidence in him.

Early in 1941, the cooperative changed its capital structure by merging the three classes of stock into a single class. The change served Urann since he still had a majority of the common stock, but it weakened Makepeace since he no longer had veto power. Also in 1941, Makepeace and Urann were charged with antitrust violations by the U.S. government. After several months of negotiation, they pleaded nolo contendere, which is equivalent to a guilty plea, and received small fines. They also agreed to give up some of their control of Ocean Spray. The new stock distribution that resulted, while still far from the one member–one vote principle of the

Exchange, was closer to the proportions of berries delivered. Urann and Makepeace still controlled about 45 percent of the common stock, but now, at least in principle, they could be outvoted.

Besides increasing its purchases of berries from the sales companies in the Exchange, Ocean Spray began to expand its membership and become a real growers' cooperative. The first new members were several growers from Wisconsin, who formed the Midwest Cranberry Cooperative within Ocean Spray. Then, during a week's trip to Washington in 1942, Urann signed up 200 growers, purchased a processing plant and other assets of a small co-op, and began construction of another plant, quickly securing the west coast for his own.

Since his membership in Ocean Spray and the Exchange didn't prohibit it, Makepeace began to process and sell dehydrated berries directly to the army at the outset of World War II. Urann also decided to sell to the army but through Ocean Spray, and he quickly opened new processing plants and retooled old ones, doubling the co-op's processing capacity between 1942 and 1944. Both Makepeace and Ocean Spray prospered at the Exchange's expense during the war. Consumer sales of fresh berries dropped because of sugar rationing, and the army wanted processed berries to feed troops in the field. In 1943 and 1944, most of the entire U.S. crop and almost all of the Massachusetts crop was dried and sent overseas, and for the first time Ocean Spray sold more berries than the Exchange.

Immediately after the war, Ocean Spray moved into the fresh fruit business, primarily because membership had increased dramatically and members were selling more of their crop through the co-op. Urann timed his move to coincide with the sharp drop in prices caused by the peacetime loss of orders. The glut that resulted in the market for fresh berries put the Exchange in a vulnerable position.

Ocean Spray and the Exchange had been uneasy allies for over 15 years: The Exchange delivered berries for canning to Ocean Spray; Ocean Spray sold some of its fruit through the Exchange; and Makepeace was a major partner in both. However, this alliance could not survive Urann selling in the fresh market. When the Exchange started to charge Urann overhead for canning his own berries, he withdrew his membership, and in retaliation the Exchange started selling berries to other canners. In 1945 Makepeace also left the Exchange for a time to market his fresh fruit independently. This meant there

were now three major opponents in the marketing war—the Exchange, Ocean Spray and Makepeace—each operating in the fresh fruit and processed fruit markets and each trying to damage the others by any means fair or foul.

The partnership of Urann and Makepeace had long been under strain. Urann was an empire builder wholly committed to processing berries. He was always quick to move, leaving the details to Quarles and others, and he often violated the spirit of cooperation by notifying Ocean Spray's officers and directors of major decisions only after he had made them. What's more, although he owned over 1200 acres of cranberry bogs, Urann was seen as a businessman and lawyer rather than a grower. Makepeace was a banker by profession, but he was a grower by temperament and tradition, used to the fresh fruit business. His family had been growing cranberries on Cape Cod and in Plymouth County since 1854 and produced more berries and owned more bogs than anyone. Urann was a newcomer and an outsider. His seemingly impetuous behavior, and especially his successes, annoyed Makepeace and the other growers.

In spite of their differences, Urann and Makepeace had more to gain from continuing their partnership. In 1948, in order to coordinate sales between Ocean Spray and Makepeace, Russell Makepeace, John's nephew and president of A.D. Makepeace Company, became vice president of Ocean Spray in charge of fresh fruit. John C. continued to operate as an independent, but he was working with Urann. Both men now clearly had the upper hand against their common rival, the Exchange.

Certain factors favored Ocean Spray over the Exchange: its financial strength, the longer shelf life of processed berries and the changing tastes of the consumers. The market for fresh berries had always been highly susceptible to gluts and sudden price drops; this was the

The Onset canning and processing plant, ca. 1940, Ocean Spray's second canning plant in the state (Hanson was the first). Located on Routes 6 and 28 in Wareham, this facility featured a cranberry information booth and a refreshment stand offering cranberry frappes, juice and ice cream. (Ocean Spray Cranberries, Inc.)

Right: A rare Ocean Spray label showing Makepeace as the canner and the packing done for the American Cranberry Exchange. (Ocean Spray Cranberries, Inc.)

main argument for cooperative marketing. Urann could drive the price down by placing berries on the fresh market and still keep most of his crop for canning. Temporarily constrained by criticism from the Exchange that he was selling on consignment—the original sin of cranberry marketing—Urann began giving fresh, higher-quality berries to the independents in an equal exchange for canning berries, knowing that his fresh berries would find their way into the market and cut the Exchange's price. The Exchange had to support prices by withdrawing excess berries from the market, either by selling them to processors or by freezing them for processing later.

Withholding berries from the market caused financial problems for the Exchange because of its weaker financial structure. Though sales were about equal for Ocean Spray and the Exchange, Ocean Spray had five times as much cash on hand as did the Exchange and the sales companies combined. Furthermore, most of the cash and other assets in the Exchange were held by the sales companies. The Exchange paid out almost all of its revenues to the sales companies, which quickly distributed the proceeds to their members.

The Exchange needed cash, but asking for more credit from the growers would be construed as a sign of failure and cause more members to switch to Ocean Spray. Moreover, Urann had a line of credit from the banks through the Cranberry Credit Corporation, which he had secured in 1942 to help finance his members.

The most important factor favoring Ocean Spray, however, was consumers' preference for processed over fresh foods. The market for canned sauce and juices was growing by about 20 percent per year. Full employment, new housing in the suburbs and the baby boom led housewives to favor foods that were easy to prepare. Regardless of the preferences of the Exchange and its growers, women were not going to continue preparing sauce from fresh cranberries in their kitchens. Either by luck or by design, Urann had been right to concentrate all his energies in processing.

In 1949 the Exchange sued for peace with Ocean Spray. The negotiated settlement was the formation of a third co-op, the Cranberry Growers Council. Both the

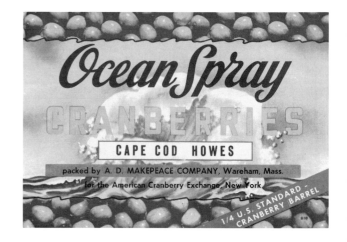

An early jar of cranberry sauce, 1930s. Ocean Spray's success throughout the '30s, '40s and '50s was a direct result of clever advertising. Cranberry products were promoted as healthful and stimulating: "Stamina without the fat," was one of the more popular slogans. On the back of this jar is the message: "Miles Standish fought, Priscilla Allen loved on a diet of cranberry sauce. It gave nerve to the warrior, sympathetic tenderness to this charming maiden. It will give you the will to do and the punch to do it." Another label reads, "A cow will give more milk, a hen will lay more eggs when happy....Easily assimilated and helps digest the other food. It puts the diner in a mood to benefit by all other food and makes possible the use of less expensive meats and fish." (Ocean Spray Cranberries, Inc.)

Left: Packing dehydrated cranberries at the Onset plant, ca. 1945. (Ocean Spray Cranberries, Inc.)

Exchange and Ocean Spray would continue to exist, but their members would belong to the council. Each would have six members on the board of directors of the new co-op, and Makepeace would have two. Fresh fruit sales would go through the Exchange and Ocean Spray would do all canning. Allocation of berries to the fresh and canning markets and advertising budgets would be determined by a two-thirds vote of the directors.

The Cranberry Growers Council did not achieve its intended affect of peace between the Exchange and Ocean Spray. Management of the Exchange wanted at first 80 percent and then 85 percent of the berries to go to the fresh fruit market. Ocean Spray wanted 40 percent for canning. The two cooperatives started going separate ways over this issue. In order to put pressure on the Exchange, Urann quietly helped start another co-op to market fresh fruit: the Cape Cod Cranberry Cooperative, under the direction of Orrin Colley, his longtime associate, and with John Quarles as legal counsel. There were now four organizations competing for berries, all selling in the fresh fruit market.

Negotiations to merge the Exchange and Ocean Spray continued over the next four years. The major differences between them were the position of Chester Chaney as director for fresh fruit sales, the rule of one member–one vote, the necessity of 100 percent contracts and the continued existence of the state sales companies. Although the Exchange was willing to give in on the first two issues, and the latter two didn't seem insurmountable, the organizations moved apart. Basically, neither trusted the other, with reason: the Exchange sold increasing amounts to private canners and Urann continued to sell fresh fruit.

The growers were caught in the cross fire. Their unsold berries were piling up in the freezers and they needed cash. They openly questioned the wisdom and usefulness of their co-ops competing instead of cooperating. Most of the attention was focused on Urann and his drive to make processing primary. A grower was either for him or against him; there was no middle ground.

161

The directors of the National Cranberry Association, 1949. Standing (left to right): John C. Makepeace, Fred Langi, Frank Crandon, Guy Potter, L.W. Resin, Harrison Goddard, Leonard Morris, Edwin W. Warners, J. Rogers Brick, Carl Urann, Russell Makepeace, Robert S. Handy. Seated: Kenneth Garside, Ellis D. Atwood, Charles L. Lewis, Marcus L. Urann, Enoch Bills, Isaac Harrison, George Cowen. (Ocean Spray Cranberries, Inc.)

Right: The directors of the American Cranberry Exchange and the New England Cranberry Sales Company, ca. 1948. Left to right: Lester Haines, Clyde McGrew, Theodore Budd, Chester Chaney, George Briggs, Arthur Benson, unknown, Vernon Goldsworthy. (Ocean Spray Cranberries, Inc.)

In 1953 the battle turned into a rout. Makepeace rejoined Ocean Spray to sell all of his crop through the co-op, and because of consumer preference, berries began to flow toward the canners. Also, Urann announced that he would buy only 20 percent of the Exchange's crop after having bought 40 percent the year before. Beaten, the Exchange offered to merge into Ocean Spray, but Urann refused, offering instead to accept only individual growers. After dragging out the inevitable, on February 23, 1954 the New England Cranberries Sales Company, the major strength of the Exchange, voted to dissolve and sell all of its assets to Ocean Spray. Its growers were welcomed as members into Ocean Spray, and although most accepted, a stalwart few refused, some waiting until Urann retired.

Selling Ocean Spray

Urann was combative by nature and no doubt enjoyed the life-or-death struggle with the Exchange. But he also firmly believed that processing rather than fresh fruit held the future for cranberries. In 1955, with the battle of the co-ops won, Urann retired; his long-time ally and antagonist, John C. Makepeace, followed two years later.

Like the founders of most industries, Urann and Makepeace were skilled in production as well as in busi-

ness—growers as well as merchants. Following them, businessmen would manage the co-op, but its members would rule it through their board of directors. Rather than sell their berries to a private cannery, growers hired experts, through the co-op, to market their fruit.

Ocean Spray was changing from a monarchy, contested by powerful feudal lords, to a form of representational democracy. And as usual in new democracies, the voters, that is, the growers, were restless. They had good reason: For several years earnings had been low and payments for berries delivered had been late. In contrast to Urann's 25-year rule as both general manager and president, the next 25 years would see six chief executive officers manage Ocean Spray.

The movement toward democracy advanced under the leadership of George Olsson. From the start, Urann and Makepeace had held enough shares to control decisions. The remaining shares were widely held among growers who delivered all of their crop, growers who delivered some of their crop and nongrowers who had received stock from Urann in return for some favor. In 1959 the co-op bought back the stock of the latter two groups and reorganized ownership so that each member held shares in proportion to the average amount of berries delivered.

The consolidation was helpful during the amino-triazole crisis of 1959, when customers refused to buy cranberry products and Ocean Spray was forced to destroy most of the crop. The future of cranberries was grim, its only hope lying in the direction of Washington and the strong support the growers had in the Republican Party. Olsson, Orrin Colley (representing the Cran-

Juice cocktail, ca. 1940 (Ocean Spray Cranberries, Inc.)

Left: Store poster, 1940s. Ocean Spray's advertising effort included billboards and posters sent to retail outlets, stressing health, economy, convenience and quality. The cocktail juice was promoted for its intrinsic health value: "A glass before breakfast is a mild laxative...Drink with meats to aid digestion...During the day relieve faintness and fatigue...Promotes sleep and a clean mouth in the morning." (Ocean Spray Cranberries, Inc.)

Bottom left: Research scientist testing cranberry juice during the 1959 cancer scare. (Life magazine photograph by Ted Polumbaum)

The Cranberry industry facing its accuser, November 19, 1959. Ocean Spray president George Olsson with Health, Education and Welfare Secretary Arthur S. Flemming after the government announced it had discovered traces of the "possibly harmful" weed-killer aminotriazole in a batch of cranberries from the Pacific Northwest. Olsson, a Plymouth attorney, fought diligently to convince the government to indemnify the growers nearly the entire value of the crop, almost $10 million. (Standard-Times photograph)

berry Institute, a lobbying group) and Marcus Urann, M.L.'s nephew, convinced President Eisenhower to order $8.5 million in restitution to the growers, which covered expenses for 1959 and kept many growers in business.

In 1972 Harold Thorkleson became president and CEO. Since Urann Ocean Spray's top managers had been salesmen, in keeping with the major function of a marketing cooperative, but they hadn't succeeded in expanding the market for cranberries. Thorkleson was a production man, but he learned marketing fast. He decided to hitch Ocean Spray's future to the public's demand for fruit drinks. Since then several cranberry–fruit blends have been successfully introduced to the public, and now over 90 percent of the crop is sold as juice.

In 1976, in a break with the tradition of not giving aid and comfort to competitors, Ocean Spray allowed a

Ike gets a bird in the hand, with plenty of garnish, 1953. Ellen Stillman (left), vice president of the National Cranberry Association and Ocean Spray's director of marketing, spearheaded a drive to sell cranberry sauce not only for the holiday turkeys, but as an accompaniment to chicken throughout the year. It's the most logical thing," said Miss Stillman, "You eat sauce with turkey, don't you? What's the difference between chicken and turkey?" The "cranberries with chicken" campaign boosted year-round sales of the sauce throughout the '40s and '50s. Cranberry displays were set up in markets alongside poultry counters, and posters were distributed promoting a new "holiday" tradition: "chicken and cranberry sauce for Father's Day." (Ocean Spray Cranberries, Inc.)

Right: Cans of cranberry sauce on sale during the "cancer scare" of 1959. (Life magazine photograph by Ted Polumbaum)

group of grapefruit growers from the Indian River area of Florida to form a co-op within the co-op. Many cranberry growers opposed this move fearing domination by the larger market for citrus fruit. In 1987 the directors voted to limit the pool of grapefruit growers in the co-op to a maximum of 15 percent.

Conclusion

Ocean Spray's recent growth has been beyond its founders' wildest dreams. One percent of 1988's $800 million in sales revenue would have bought the entire cranberry crop in 1930, the co-op's first year of operation. In 1985 Ocean Spray broke into the Fortune 500 and now ranks 385. Sales are expected to top $1 billion within the next few years.

Some of this growth has come from the increase in the annual harvest—the first 1-million-barrel harvest was in 1953, and now about 4 million barrels are grown in the United States—but most of the growth is from higher prices for cranberry products. Constantly rising prices in spite of bumper crops are a remarkable achievement in marketing farm produce.

Obviously, a meager crop hurts the farmer, but bumper crops usually don't help either; selling perishable crops in a limited market can cause ruin as surely as crop failure can. Cranberry growers have usually been spared this predicament, particularly in recent years when prices

have risen as the harvest has grown.

Consumer preference for fruit drinks and the general concern over health have certainly helped sales at Ocean Spray, but the cooperative form of business organization has been the main reason for the success of the cranberry industry. Economic studies have shown that an agricultural marketing co-op can be successful if it is well managed, if it has the commitment of its members and if it can control a large share of the market. Ocean Spray has managed to do all three.

Both Ocean Spray and the American Cranberry Exchange were fortunate in the quality of their management. Chaney, Urann and Makepeace put the industry on solid ground from the start. Olsson, Thorkleson and Gelsthorpe understood the importance of political connections, efficient production techniques and communication between growers and processors. All of them paid great attention to creative though dependable marketing, making Ocean Spray a leader in new product promotion and innovative product design.

From the start, most cranberry growers, especially those with small holdings, were committed to the co-op. Marketing their own crop was expensive and they needed protection from unscrupulous buyers. The early success of the Exchange confirmed their commitment, and the co-op became an article of faith. Larger growers took more convincing. John C. Makepeace mistrusted

the co-op in the beginning and refused to join because he had much to lose as the largest grower in the world. Yet in 1914, three years after he joined, he stated,

> It means the substitution of intelligence and the instruments of modern commerce for sheer bull-headedness and obstinacy. No grower or dealer on the outside is decrying or belittling the work of this organization except for selfish motives or from ignorance.

Though committed to the co-op by contract and belief, members have never been passive followers but have always actively participated in decision making. Unpredictable and powerful, nature and the market have made farmers a cautious lot, reluctant to change and suspicious of great promise. Active participation (often in heated argument), coupled with full commitment, strengthened the co-op's bonds.

In spite of his 50 years as a dominant figure in the Exchange and Ocean Spray, Urann never gained the trust of many of the growers. Even now, 25 years after his death, mention of his name stirs controversy. Nevertheless, if they could be convinced that his strategy or plan was correct—usually after a long struggle—the members gave him their support.

Currently Ocean Spray markets about 85 percent of the total U.S. cranberry crop, and every growing area is strongly represented. Management has always aggressively pursued market share, but by limiting supply, nature has had the final say in how big a share the co-op will get. Urann said it best:

> We are fortunate in that the area in which cranberries can be grown is limited. This places a natural restriction on over-production. It also confines growers to small areas where they can become acquainted with each other. The cranberry industry is probably outstanding for its friendly feeling and lack of competition among growers. This has played a great part in furthering cooperation.

Waving fields of golden grain and vast orchards of succulent fruit, easy to grow and harvest, have defeated all efforts of farmers to cooperate in the market. Cranberries, grown in inhospitable surroundings and too bitter to eat off the vine, have produced a small number of growers who had no other choice but to work together.

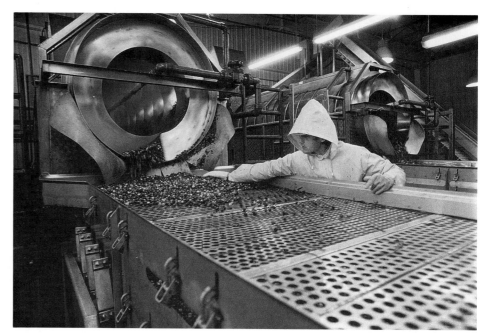

A truckload of berries being tested at the Ocean Spray receiving station, South Carver, 1989. Wet-harvested berries are checked for trash content (debris and chaff) as well as for color, density and pesticide level. Incentives are offered for good color quality and purity. A hose randomly sucks out a sampling of berries from the load, tests it and applies the results to the entire load. The trash content is subtracted to determine net weight. (John Robson photograph)

Right: The screening area at Ocean Spray's receiving station, South Carver, 1989. The berries are run through a "barrel washer" (top center) and a stick remover and are then sent along a conveyor that grades them by allowing the smaller ones to fall through a grate. A vibrating action moves the fruit forward. At the end of the conveyor, two brushers, 9/32-inch apart, turn in opposite directions, forcing the soft berries through the brushes and into a bin. (Joseph D. Thomas photograph)

Unloading berries at the South Carver receiving station, 1989 (John Robson photograph)

Beaton Growers' Service workers on Ocean Spray's Poquoy Bog in Lakeville, removing floating debris after the harvest, 1989. In the background is the company's corporate headquarters. (Joseph D. Thomas photograph)

All in a Package

Michael Zaritt

Neither sweet like apples nor tangy like oranges, the cranberry could never be plucked from the vine and eaten raw. Even in the heyday of fresh fruit sales, the berry needed a lot of cooking and a lot of sugar before it could be served with the holiday turkey.

Over the years, packagers have tried many innovative ways to bring cranberries to the dinner table. They have brought us juice and cocktail, ketchup and mousse, dried fruit and sherbet. Some products have succeeded and others have failed.

Selling the cranberry has always been a marketing challenge, but it is one that the industry has faced with skill and imagination.

An advertisement, ca. 1940. When the juice cocktail was introduced in the early 1930s, it was immediately touted for its health value, and as this promotional display card read, it was "Especially important for those suffering from indigestion, gastric disorders, and for invalids and convalescents whose digestive systems have become sluggish." (Ocean Spray Cranberries, Inc.)

Left: Chicken and cranberry sauce for Father's Day, 1953. (Ocean Spray Cranberries, Inc.)

Henry VIII, digging in for a British advertising campaign promoting chicken and cranberry sauce, 1950s. (Ocean Spray Cranberries, Inc.)

Opposite page, top: The first bottled cranberry juice cocktail, 1930s. (Ocean Spray Cranberries, Inc.)

Opposite page, bottom: The information and retail sales stand at Ocean Spray's Onset plant. According to an internal memorandum to growers, the stand "provides advertising as well as a substantial profit. Fifty thousand visitors stop annually to buy cranberry frappes, Cape Cod Sundaes, and Ocean Spray Cranberry Sauce, Cocktail..." (Ocean Spray Cranberries, Inc.)

Radio Cranberry, 1940s. M.L. Urann did a series of radio broadcasts about cranberries and the cranberry industry. Here he interviews Ocean Spray directors. Left to right, Russell Makepeace, G. Mann, M.L. Urann, Charles Lewis, Isaac Harrison. (Ocean Spray Cranberries, Inc.)

Top right: On the Today Show, 1955. Seen with Dave Garroway (left) are "Cranberry Girl" Diane Hilliard, "Cranberry Boy" Ken Nye, and Miss America, Lee Meriwether (right). The cranberry boy and girl were selected at the Cranberry Festival held annually at Edaville Railroad and sponsored by the National Cranberry Association. (Ocean Spray Cranberries, Inc.)

Arthur Godfrey, getting his hands dirty. Godfrey's enthusiasm for cranberries stemmed from his belief that a diet including plenty of cranberry juice helped cure him of cancer. In 1964 he invested money in a Cape Cod bog and for a time was a member of the Cape Cod Cranberry Growers' Association. His ad-libbed TV and radio commercials for cranberry juice were an important part of Ocean Spray's national media campaign in the 1950s and '60s. (Ocean Spray Cranberries, Inc.)

Cranberries for lunch at the Mayflower Elementary School in Middleboro, 1960s. Under George Olsson, Ocean Spray sold 10 million pounds of cranberries to the Department of Agriculture for school lunch programs in 50 states. (Ocean Spray Cranberries, Inc.)

Bottom left: Stringing cranberries, a Cape Cod tradition, 1960. No one knows exactly when stringing cranberries as Christmas decorations began. One elderly grower says, "My father did it when he was a child in the 1870s and it probably goes back further than that." (Ocean Spray Cranberries, Inc.)

Betty R. Bowen of Wareham High School, leading the 1960 Cranberry Easter Parade. (Ocean Spray Cranberries, Inc.)

171

Ellis D. Atwood and Elthea Atwood in front of a train he named for her, 1940s. Ellis D. Atwood was a very successful cranberry grower. At one time he held the largest privately owned bog in the world, 210 acres. To those who knew him, however, he was a generous man who cared about his community. He built a baseball field with night lights on his property and provided the Carver team with transportation and equipment. When he died in 1950 he was working with M.L. Urann and John C. Makepeace to organize Ocean Spray. (Courtesy of Larry Cole)

Below: The passengers onboard an Edaville train see the harvesters at work, ca. 1955. Ellis D. Atwood was a train buff. In the 1940s he acquired Maine's last working 2-foot-gauge railroad, including 5-1/2 miles of track, and set it up to haul sand and berries on a route he surveyed himself through his bogs. The train caused quite a stir among his neighbors throughout the '50s, so Atwood added a few passenger cars for rides and soon built a station and hired former engineers and conductors to operate the railroad on a regular schedule. This was the beginning of Edaville Railroad, which in its time has carried well over ten million passengers. The name "Edaville" (short for E.D.A. village) originally referred to the village of about a dozen bog shanties and screenhouses on the Ellis D. Atwood bogs. Ocean Spray Cranberries, Inc.)

An Edaville Railroad display, ca. 1958. Soon after the railroad started operating, Atwood and his wife Elthea decided to move their annual Christmas display of lights and moving figures from behind their house on Sampson's Pond to various points along the train route. Thus began Edaville's holiday tradition. (Ocean Spray Cranberries, Inc.)

Bottom left: The Cranberry Festival at Edaville, South Carver, 1948. Charles Rhyner and Teeny Wilson do the "cranberry bounce" for Marcus L. Urann (seated left) and cranberry queen Marcia Williams (background right). Every year for over 50 years, Massachusetts cranberry growers celebrate their harvest with a cranberry festival to which the public is invited. (Ocean Spray Cranberries, Inc.)

The crowning of the Cranberry Queen at the Cranberry Festival, ca. 1956. The queen appeared on behalf of the industry at promotional events, fairs, exhibits, before groups, on television and radio, etc. (Ocean Spray Cranberries, Inc.)

Part Four

Cranberry Growing Today

Overview

Christy Lowrance

Land use and water quality have become critical issues in southeastern Massachusetts, particularly for cranberry growers. Because of the massive development that has taken place within the last 15 years, open land is dwindling at an alarming rate, and the increase in activity and population has caused contamination of water supplies.

Many of the cranberry growing areas on the Cape and in Plymouth County rely on an underground water source, or sole source aquifer. This means that contamination in one part of the aquifer has the potential to affect the entire water supply. Any risk to the water by chemical application on land is thus a concern to abuttors and environmentalists. It is also of great concern to the grower:

> If we found one [chemical] that might enter the ground water, we wouldn't use it. Anything that happens out there could affect my family because we live adjacent to my bog…We sample and check all our pesticides. The crops are checked by the USDA…(Doug Beaton)

The cranberry industry is quick to point out that large bogs protect water supplies because of the filtering and storage capacity of their accompanying wetlands.

Mutual interests have stimulated cooperation between the industry and environmental groups. In 1988 the Cape Cod Cranberry Growers' Association and the Massachusetts Audubon Society co-sponsored a pesticide bill that passed in the state legislature by a vote of 154 to 0.

Besides demanding hours, environmental concerns, insect pests and poor weather, growers today have another common problem: the public.

> People don't have any idea what we're doing. They just think the bog is a perfect place to play softball and drive golf balls, and they hate chemicals and helicopters. There is a state law that you have to post bogs when pesticides are being used, but it makes people nervous even if you do it at night and it all washes off by morning. (Linc Thacher)

> Two summers ago, my people shooed children off a bog and their parents turned around and sued us. Then the judge said we were criminally liable. We've tried no trespassing signs, but they don't stay. During harvest time, people like to harvest their own. (Arthur Handy)

Leadership and organization in both the nineteenth and twentieth centuries have enabled the industry to meet many challenges. But a strong sense of family tradition has also helped.

> You have to have an understanding family, especially in the fall, when you are working seven days a week, sometimes 24 hours a day. If there is going to be a frost, you put your wedding anniversary celebration on hold. From our family's point of view, the cranberry industry is a lifestyle. (Doug Beaton)

At a time when land in southeastern Massachusetts is worth remarkable figures, why do growers still put in long, hard days, watch thermometers at four in the morning and pacify irate neighbors to make a living? "If you were talking with the wife, she'd say it was for the birds because it is such a struggle to show a profit," one grower admitted. But for others, the satisfaction of being a cranberry grower is not measured just by the market value of a barrel. "This is where we get our pride," says Doug Beaton, "It's us being able to leave the bogs for future generations as good as we received them. I look on cranberries as our heritage."

Putting down herbicides. The white foam markings show the driver where he has already been so that he won't apply more herbicide to the same spot. (John Robson photograph)

Opposite page: East Head Reservoir, Miles Standish State Forest, South Carver. East Head Reservoir is the source of the Wankinco River and feeds hundreds of acres of cranberry bogs—most belonging to the A.D. Makepeace Company (Wankinco Bogs). Here, the reservoir diverges into two streams. To the left, the water flows for Henry Davison's East Head Bogs, and to the right, for Makepeace. The growers are allowed to take water to the level set at the bottom of their flumes. State law restricts their taking water below a certain level. Davison, because he holds original water rights, is allowed to take a couple of inches lower. (John Robson photograph)

Stewards of the Wetlands

Linda Rinta

Cranberry growers own and manage over 61,000 acres of open space in Massachusetts. Of these about 12,000 acres are bogs and the rest is surrounding woodlands, wetlands and brush. In a region where forests and undeveloped land are quickly disappearing, this acreage represents much of what little open space is left.

The open space held by growers provides refuge for all types of wildlife, including the endangered red-belly turtle, and rare and endangered wild flowers and plants such as orchids, holly and mayflowers (the state flower). Growers host Massachusetts Division of Fisheries and Wildlife bird nesting boxes, and many plant trees, create bee pastures and construct bat roosts.

Because cranberry production is effective in providing open space, conservation commissions in many towns, including Chatham, Carlisle, Dennis, Falmouth, Hanover, Harwich, Mashpee, Wareham and Yarmouth, manage commercial bogs. On Nantucket cranberries support a 1300-acre reserve.

Cranberry growers think of themselves as stewards of the wetlands. They believe they have husbanded the watersheds better than any other group—even better than state and local protective regulations—and they point to their reservoirs as evidence. Cranberry production depends on abundant clean water, and the bogs and their uplands require a whole system of man-made reservoirs, canals, ditches and flumes to provide it. In Plymouth County alone, growers maintain 4611 acres of water storage area, which accounts for 22 percent of the entire surface water inventory. During the annual winter flood, an additional 12,000 acres of stored water recharges the local water table.

In wetland bog systems, flood water is recycled as it

passes from bog to bog through canals and flume gates. Exiting water is captured in holding ponds and reused, often shared by several growers. These reservoirs create boundary wetlands that nurture animal life, filter storm runoff, collect sediments and control flooding. In general, wetland cranberry bogs constitute some of the safest soils to farm, because their clay base allows the movement of water within a perched water table, protecting the aquifer below from exposure to agricultural chemicals.

However, wetland bog construction does mean the loss of trees and bushes that serve as habitats, nesting sites and protective cover. The Army Corps of Engineers and the Massachusetts Wetland Protection Act regulate wetland bog building and can deny growers permission to carry out such projects.

Today most new bogs are constructed in uplands, which lack the natural soil structure that supports cranberry vines—clay, peat and sand steeped in surface water. Particularly susceptible to drought, upland bogs require constant attention to produce even modest crops, and worse, they lack the impervious liner of clay that protects the ground water underneath. Unlike wetland bogs, where the perched water can be filtered, stored, reused and monitored, water moving through sand bogs in the uplands is quickly lost to the aquifer.

Construction of upland bogs is expensive, entailing high engineering, permitting and insurance costs. The biggest hurdle to upland bog expansion, however, is spiraling land values and property taxes throughout the region. The capital investment needed to cover enormous mortgages and equipment costs can strap established growers, and it is especially hard on individuals hoping to break into farming. For most growers, whether bog owners or tenant farmers, there is little margin for error: A single poor crop can be devastating.

On average, growers own four acres of surrounding land, including wetlands and reservoirs, for every acre of producing bog. This land serves as a necessary buffer zone between bogs and their neighbors, but it is becoming harder to hold on to in Massachusetts' competitive land market, where the land has more value for development than for agriculture.

Cranberry growing depends on unpolluted watersheds, which are being violated by nonpoint source pollution such as seeping septic systems, road runoff and oozing landfills. This problem is exacerbated by increasing population and development throughout the region.

Many neighborhoods have grown up bogside, replacing

the surrounding open space. New neighbors may at first be attracted by the beauty of the bogs, but without a protective buffer zone, all-night activity during the two yearly frost seasons, noisy truck and helicopter activity and pesticide spraying eventually lead to antagonism between residents and growers.

Children are attracted to water, but ditches, water holes, flumes and pumps are unsafe playgrounds. Even adults don't understand that these are work areas, and they often cause damage through accident or misunderstanding. One morning, for example, a grower arrived at his bog to find his sprinklers shut off and a note stuck to his pump that read, "Your pump was keeping me up. Please do your watering during the daytime hours." The sleepless neighbor, bold enough to sign his name, had no idea he had cost the farmer almost half his yearly income from frost damage.

The steady appearance of houses, condominiums and trailer parks has transformed cranberry towns like Carver into Boston suburbs. As these communities lose their rural character, local regulators become insensitive to the

Applying herbicide on a Griffith Company bog in South Carver. Mike Sylvia of Cranberry Consultants, Inc. says the company specializes in insect and weed control using Integrated Pest Management (IPM) techniques as well as chemical applications. (John Robson photograph)

Opposite Page: A Rochester cranberry bog. (John Robson photograph)

interests of agriculture, and longstanding farming practices are threatened.

For instance noise pollution bylaws decide the time and type of equipment used, and some neighborhoods have banned truck traffic. In Falmouth fertilizer application by helicopter is banned before 8 A.M., even though application during the early hours would be safer for the public. In some towns growers are required to notify state and town officials five days prior to applying a chemical, undermining the daily monitoring for insect damage that would reduce the number of sprays.

Farming anywhere in the United States is no longer simple, but cranberry growers in Massachusetts are determined to hang in for the long term, as one grower said, "like a 200–pound snapping turtle!" To do so, they must update their machinery as well as their ideas and skills. It is no longer enough simply to grow a crop well.

The environmental challenges of farming represent problems of economics. In order to preserve protective

Sprinkler systems, Morse Swamp Bog, Wareham. (John Robson photograph)

Sanding on Makepeace Company bogs. Foreman Ricky Kiernan says the sanding trucks are made in the company shops from scrap airplane tires and Briggs & Stratton two-cylinder engines taken from old trucks. The winter of 1989 was unusually mild and almost no ice sanding was done. Growers did their winter sanding with bog buggies, by helicopter or with wheelbarrow and shovel. (Joseph D. Thomas)

cranberry bog upland, growers must manage their natural resources with a shrewd business sense and steely resolve.

The Lesson of Log Swamp

Al Pappi was the master of Log Swamp for many years. The 90-plus acres of bog and 500 acres of surrounding woodland and swamp were his pride. Nevertheless, when he died, the estate taxes made the sale of Log Swamp a financial necessity for his heirs. At auction in 1988, the selling price was determined by the value of the existing bogs, the surrounding gravel and timber resources and the space for potential new bogs.

The existing crop could not support the swamp's several-million-dollar price tag, so to pay for the property, the new owners lumbered the high pine knolls and mined the gravel. The neighbors were outraged at the destruction. Even though the resulting moonscape was replaced with new cranberry bogs and the area is once again beautiful, it is no longer Log Swamp.

The story of Log Swamp raises many questions: What is the effect on local drainage and aquifer recharge when a high, permeable place is made into a low place? Are gravel-rich towns in danger of being over-mined? Will the

West Barnstable grower Jim Jenkins, walking through the abandoned No Bottom Pond Bogs adjacent to his family's bogs. "It's been 25 years since they've been picked....Construction and tourism have replaced cranberry growing on the Cape." The town of Barnstable will not allow Jenkins to restore the abandoned sections of the old bog that belong to him. The town says it has to be a producing (i.e., commercially harvested) bog before it can be restored—the wild berries that grow in abundance don't count. This bog will revert back to its original wetland condition. Some Cape towns, such as Yarmouth, Mashpee and Hyannis, have bought cranberry bogs and turned them into revenue producers. (John Robson photograph)

A cedar swamp off Highland Road in Lakeville is one source of water for John Egger's bogs. (John Robson photograph)

179

high cost of land drive growers to exploit their uplands to a point of diminishing return, and will this lead to the demise of the cranberry industry?

Log Swamp also raises the question of what is the best use for the land. What is the value of conservation? Maybe those who conserve water, land and resources should be compensated in some way for their efforts. In Massachusetts conservation and cranberry growing are interdependent. As one goes, so goes the other.

Growing Berries Under the Stack

When the SEMASS trash-to-energy plant was first proposed in Rochester, many cranberry growers feared the environmental problems it might cause. There are about 1200 acres of bog within a 3-mile radius of the plant, and fragile watersheds in that area feed many other bog systems. What gas would be brewed in such a plant? What would be the effect of fly ash on the region? How would

SEMASS, Rochester. (John Robson photograph)

the waste be handled? What about the smoke?

As manager of the Slocum-Gibbs Cranberry Company, Gary Garretson oversees a 500-acre piece of property including bogs, watersheds and timberland abutting SE-MASS. At the first sound of construction equipment, he set about educating himself on stacks, emissions, scrubbers, soils and the hydrology of the area. He also made sure that the environmental concerns of cranberry growing became the concerns of the SEMASS engineers.

It was a huge investment of time and money, but Garretson's efforts paid off. Today he is satisfied with SE-MASS's environmental efforts. The plant's steam is generated, recaptured and reused within a "closed system"; fly ash is carefully stored and transported in sealed cells to protect the surface and ground water; and all effluent is collected and treated. Also, SEMASS is the first plant in Massachusetts to have state-mandated dioxin limits.

As a vigilant protector of his property, Garretson will continue to monitor the soil, air and water on Slocum-Gibbs property for any signs of sick and failing vines. Most important, he has established a dialogue with the management and engineers at SEMASS.

Industrial neighbors are not new to cranberries. In colonial and early industrial times, bogs shared mill ponds created by the iron industry, and some of the earliest Plymouth County bogs grew from the wastelands of colonial strip mining. Cranberries also shared the industrial labor pool. In the late nineteenth century, much of the industry's workforce came from the New Bedford textile mills. Many workers left the noise and damp, closed air of the cotton mills for the fresh air of the bogs.

But unlike most types of manufacturing, cranberry growing depends on the use and reuse of clean water. As a result of residential development, however, the water in southeastern Massachusetts has become increasingly vulnerable to nonpoint source pollution, the single most degrading factor in the environment.

From the growers' perspective, no urban encroachment is welcome, but the worst-case scenario is a 100-unit housing development. Light industry, controlled and monitored, is not viewed with such dread because it is required by law and public outcry to be environmentally responsible and can be held accountable, as homeowners rarely are, for lapses.

As southeastern Massachusetts continues to grow in population, cranberry growers will need to play a greater role in community planning. Following the example of Gary Garretson and the Slocum-Gibbs Cranberry Company, they need to stay well-informed to better protect their resources and environment.

Some steps have already been taken in the direction of environmental leadership. In Carver, growers played a significant role in drafting and passing an aquifer protection bylaw, and they are active in creating a gravel removal bylaw regulating bog sanding, expansion and gravel mining. At the state level, the Cape Cod Cranberry Growers' Association worked with the Audubon Society and other groups on a law ensuring responsible use of pesticide chemicals.

The growers' association routinely conducts safety training sessions for cranberry workers, and they have pioneered Integrated Pest Management (IPM), which promotes nonchemical cultural practices such as bog sanding and dike burning to eradicate pests.

More and more growers are looking towards natural predators for pest control. Biological warfare sounds like the stuff of science fiction, but recent research has discovered fungi that attack insects, bugs that eat other bugs and viruses that kill only pest insects. Growers provide funding and the use of their bogs for much of the research, which is being carried out at the University of Massachusetts Cranberry Experiment Station and at the Universities of Wisconsin, Washington and Florida, as well as in laboratories at Tufts University, Rutgers University and Ocean Spray.

The New Alchemy Institute in Falmouth has worked with growers on parasitic wasps and nematodes since 1985. The prospects for the nematode in pest control look good. Another hopeful is the virus B.t. Dipel, which is now being used experimentally on the bogs.

Many growers demonstrate their commitment to environmental issues by serving on conservation commissions throughout the growing area. Every cranberry town has grower commission members with a history of public service. Self-serving, some say. Indeed. Cranberry growers have a keen interest in the welfare of the environment; their trade depends on it.

The Massachusetts Cranberry Experiment Station and the State Bog, East Wareham. (John Robson photograph)

Rochester grower Kirby Gilmore, pondering the reconstruction of a flume gate destroyed by the previous night's high winds and flood waters. (John Robson photograph)

181

On the Bogs

John Robson
Joseph D. Thomas
Beverly Conley

Pete Mason, installing sprinkler heads at Ralph Thompson's bog off Great Neck Road in Wareham. Federal officials will inspect the system and give Thompson a monetary credit for his conservation practices. (John Robson photograph)

Top left: Pitching cuttings, to be set by a truck-rigged vine-setter on Joseph Barboza's Morse Swamp bogs in Wareham. Early growers debated the proper distance between groups of newly-planted vines. Today's methods simply call for an even distribution. (Joseph D. Thomas photograph)

Spreading sand in preparation for planting on a rebuilt section of Makepeace Company's Morse Swamp Bogs. The workers will spread vines by hand onto the bog surface and harrow them into the ground with the mechanical vine-setter at left. Bogs are sometimes rebuilt if the vines become too thick and difficult to harvest or, as is the case here, if the surface is so out of level that flooding the bogs to a uniform depth is impossible. (John Robson photograph)

Opposite page: A spring blossom on Ted Young's five-year old Carver bog. (John Robson photograph)

Filling the ends of cross ditches with sand, Harju bog, Middleboro. Ronnie Harju (right) throws a shovelful of sand toward Eddie Harju, who packs it into the ditch. Filling the ditch allows the water-reel harvesters to cross from the shore to the bog without planks. (John Robson photograph).

A grower sanding the old-fashioned way—with wheelbarrow and shovel. The deep, triangular frame of the barrow extends beyond the front of the wheel, giving it better balance when lifted. (John Robson photograph)

Gary Florindo, releasing sand from the hopper of his sanding buggy after planting at Morse Swamp. (Joseph D. Thomas photograph)

David Mendes, forklifting bees at 13 Acre Bog in South Carver. (John Robson photograph)

Top left: Victor Mercado, trimming weeds at Frogfoot Bog, South Carver. (John Robson photograph)

Mud lifting on Frogfoot Bog. Mud and weeds cleaned from the ditch are placed on mats and airlifted to the shore. (John Robson photograph)

An aqueduct carrying water over the trestle at Maple Springs Brook, Wareham. Below the trestle, the brook winds its way through the Maple Springs bogs, built by A.D. Makepeace around 1900, forming sharp contours and odd shapes that characterize Massachusetts bogs. (John Robson photograph)

Laying irrigation piping at Benson Pond Bog, Middleboro. Darren Morris and Ben Gilmore attach the flexible irrigation piping to their tractor, which will cut an 8-inch-deep furrow into the ground and pull the pipe through as it drives along. Kirby Gilmore's restored 1934 Ford truck and 1946 Ford tractor are evidence of growers' resourcefulness in keeping their machines productive. (Joseph D. Thomas photograph)

Installing a backflow preventer, or chemigation valve, on a spray point pump at A.D. Makepeace's Big Bog, along the Wankinco. Ken Knight (welder) has been welding for cranberry growers for five years, and 90 percent of his work is on irrigation systems. The valve he is installing is an EPA requirement; it prevents chemicals fed into the water being pumped onto the bog from going back to the water source if the pump shuts down. Ricky Kiernan points out that Makepeace Company pumps have two valves that automatically shut off when the pump shuts down, thus forcing water to return to the source untouched by chemicals, and he thinks the valve is useless. Ken knows a couple of people who do not have or do not use the option that closes the valve. He thinks the EPA developed the law in response to problems with large farmers out West who were drawing water from wells that supplied drinking water. He has never seen such a problem in his work around here. (John Robson photograph)

Top left: Water flowing onto David Olson's bog in Carver. (John Robson photograph)

Flooding a bog with a tractor-driven pump, Harwich. At the Exit 10 Bog at Pleasant Lake, growers from the Mello-Wilson Cranberry Company take the water from one bog and pump it into another. (John Robson photograph)

187

Ted Young, harvesting with a Furford machine, Carver. Young built and planted this bog five years ago. This is his first big crop. (Joseph D. Thomas photograph)

Dry-harvesting on New Piece Bog, Carver. "Biggest berries I've ever seen," remarked Gertrude Rinne. Her nephews, Paul and Larry Harju, harvest the bog for Gertrude and her husband Oiva. They own three acres and do all dry picking. (John Robson photograph)

Scooping berries along the bog ditch, Stuart Bog. Hand-scooping berries still has its place on many dry-harvested bogs. Harry Stopka of Rochester, who has worked for Decas Cranberry Company since he was a boy in the 1930s, pauses before crouching down along the ditch to scoop berries that the Furford pickers were unable to reach. (John Robson photograph)

New Piece Bog, Carver. Workers from Puerto Rico and Scotland supply today's cranberry industry with most of its labor force. Most of the workers from Puerto Rico reside in New Bedford and work from March through December on all aspects of bog maintenance and harvesting. Here workers tie large nylon sacks, called berry bags, to the hoop frame that holds the sack open. The burlap bags on the ground are filled by workers operating the Furford picking machines (left) and deposited in rows as they push along. Other workers then dump the bags into the hoop-framed berry bags. When filled these bags are lifted by helicopter to the shore and dumped into large bins, which are loaded onto a truck and hauled away. Richard and Paul Harju developed the hoop and berry bag system of helicopter lifting. "In the old days," says Paul, " it took a lot of time to load a truck. You'd go to Ocean Spray, put boxes in the truck one by one; unload them at the bog one by one; after they were filled, reload one by one; and then unload at Ocean Spray one by one." (John Robson photograph)

Darlington harvesting at Griffith bogs, South Carver. On Clark Griffith's #22 Bog, formerly the Marc Atwood Bog, only a half-acre is dry-picked, and only because its level is too high to water-pick. At the first opportunity, Griffith will rebuild the bog to level it with the adjacent bogs. Today's crew of nine (four full-time, year-round workers) work amid the menacing briars, especially the silverleaf, that constantly jam up the machines. "They're out of control," laments Clark. It will take the crew over four hours to pick. "Every bit of trouble we could have, we're gonna have today," he says as he crawls under a Darlington to fix its dented skid plate. (Joseph D. Thomas photograph)

Lifting berries on Benjamin Bog, South Carver. John Clark operates his helicopter with the dexterity and precision of a surgeon. "It beats working for a living." (Joseph D. Thomas photograph)

Removing the hoop frame from the berry filled bag, ready for helicopter lifting, New Piece Bog, Carver. (Beverly Conley photograph)

A young worker from Puerto Rico and his Furford harvesting machine, Stuart Bog, Rochester. (John Robson photograph)

Marshall McCarty, dumping berries into a cranberry bin to be airlifted, Stuart Bog, Rochester. (Joseph D. Thomas photograph)

Lifting berries off Stuart Bog, Rochester. Only Furfords are used for harvesting these bogs. Bill Chamberlain, foreman for Decas Cranberry Company, said "The best day I ever had with Darlington pickers was taking off 600 barrels in 1980. With six Furford machines, I have picked 110 of these 10-barrel bins. We can do with six Furfords what we did with nine Darlingtons. My father had the record. He took off 125 bins in one day." (Joseph D. Thomas photograph)

The Decas Cranberry Company got its start in 1935. William, Charles and Nick Decas sold and distributed fresh fruit. Their market on Main Street in Wareham was across from the home of well-known grower and banker, Leck Handy. Handy visited Bill Decas one day and told him that the bank was foreclosing on a bog near Mary's Pond in Rochester and that he thought it would be a good investment for the Decas brothers. Bill Decas, who had a high regard for Handy, said "If Mr. Handy thinks it's a good investment, then it must be good," and he bought the 15-acre bog.

Today, Decas Cranberry Company is the oldest and largest independent cranberry distributor in the country. It is also the largest dry-harvesting grower, with 450 acres of bogs. This year Decas will handle 250,000 barrels of the 400,000 grown by independents. It buys and packages cranberries for other large independents, such as PALS (Peter A. LeSage Co.) and Hiller Cranberries of Rochester, and sells them under the Paradise Meadow label. (Joseph D. Thomas photograph)

The interior of a Morse Bros. Cranberry Company shed, South Middleboro. Years ago separating and screening were conducted here; today the old screenhouse is a storage shed for equipment and bags of cranberries that did not get lifted off the bog. (Joseph D. Thomas photograph)

An old flume gate on an A.D. Makepeace Company bog in Carver. (Joseph D. Thomas photograph)

Mud lifting, Rochester. In the winter, the dormant cranberry vines take on a vibrant dark maroon that contrasts sharply with the stark winter whites of the nearby sand dunes and icy waters. Winter is not a dormant time for cranberry growers. Sand must be spread, ditches cleaned, irrigation systems installed and other general maintenance work performed to ensure a healthy crop in autumn. (Joseph D. Thomas photograph)

Spring rain, ferns and a pump house on the Wankinco River, near Big ADM Reservoir at Wankinco Bogs. (John Robson photograph)

193

Laying "black poly" flexible pipe for a sprinkler system, Benson Pond Bog, Middleboro. (Joseph D. Thomas photograph)

A springtime greening up, Big Bog, Wankinco. (John Robson photograph)

194

Flowers blooming at Ted Young's bog along Route 58 in Carver. (John Robson photograph)

Top left: Mud mats ready for lifting at Frogfoot Bog near Tihonet. (Joseph D. Thomas photograph)

At Big Bog in Wankinco, plants, reptiles and small varmits find lush corners like this a welcome habitat. The only thing is, these ditches eventually must be cleaned to allow proper water flow. The yellow weed covering the corner of the vine is the dreaded parasitic crawler known as the dodder weed, which robs nourishment from the cranberry plant. (John Robson photograph)

Susan Makepeace and Judy Rounseville, aboard the A.D. Makepeace Company float at Wareham's 250th anniversary parade. (John Robson photograph)

Top right: George Peck Jr. of the North Carver Cranberry Co., running a weed wiper over his Goudreau Bog in North Carver. He is applying a herbicide to the leaves of the weeds. George bought the bog three years ago. It was overrun with weeds at the time and he is trying to bring it back. "You couldn't see the vines. The neighbors keep saying 'Looking good, George' when they pass by, so I guess it's getting better." (John Robson photograph)

Field boxes laid out in anticipation of the first dry harvest, old Ben Shaw Bog, South Carver. In the background is a sand and gravel mining pit—a burgeoning Carver industry. (John Robson photograph)

Gay Head Wampanoag children handpicking on Cranberry Day at Lobsterville, Gay Head, Martha's Vineyard. (John Robson photograph)

Berries ready for picking. (Joseph D. Thomas photograph)

Hand-scooping along the bog ditches at Stuart Bog in Rochester. Jeff Thomas from Indiana scoops along with his father-in-law, Henry Stopka of Rochester, for extra cash and for the experience. He is a mechanic for a truck leasing company and admits that scooping is hard work. "It's more difficult than working on a soybean farm back home." (Joseph D. Thomas photograph)

Susan Mann of Plymouth, corralling berries at the Mann family's Garland Bog along Interstate 495. These berries are a large variety called Stevens, and their red and white color create a swirling pattern as they float in the water. The Stevens berry is from Wisconsin; it generally has a high yield and, if harvested late in the season, turns a deep red. (Joseph D. Thomas photograph)

An idle water reel at Chipaway Bog, East Freetown. Grower Steve Ashley builds his water-reel harvesters, as well as his other machinery, in his garage. "I started with a Chevy rear-end, leveled it out on the floor and started welding it together from the bottom up. I work on the design as I go along. I'll think of changes at night when I go to bed and then apply them the next day." (John Robson photograph)

Richard Harju, with a Furford machine at New Piece Bog, Carver. Richard and Paul Harju manage this bog for Gertrude and Oiva Rinne. They hire a six-man crew from Scotland who work from April until December. "The government sets the rate, we pay the transportation," says Paul.

Lifting berries at the Stuart Bog, Rochester. John C. Decas of Decas Cranberry Company says his company was the first to use helicopters to lift cranberries. In 1978 he took the idea to a friend, pilot John Clark, and they began experimenting. They tried using open skids, tomato crates and bags, finally settling on 10-barrel. wooden bins. Decas found that berry bags, used by some growers, have a tendency to crush some of the load when the drawstrings tighten and the bags are lifted.

The advantages of helicopter lifting are many: Bog damage from the bog buggies is eliminated; roads on the bog made by buggies are replanted; the work is less arduous; the workday extended (helicopters can work in the dark); berry spillage is reduced; manpower is reduced; and more than twice the amount of berries can be taken off the bog in a day. (John Robson photograph)

Water-reel harvesting, Poquoy Brook Bog at Ocean Spray headquarters in Lakeville. Ocean Spray's bogs are maintained by Beaton's Cranberry Growers' Service of West Wareham. This is the last day of harvesting. Workers will work until dark and take the berries off the bog under floodlights. (Joseph D. Thomas photograph)

Water-harvesting in Carver. The berries have been corralled by the wooden booms and are being pulled toward one end of the bog where they will be taken out. (Joseph D. Thomas photograph)

Dumping berries into berry bags to be lifted by helicopter, Morse Bros. bog, South Middleboro. (Joseph D. Thomas photograph)

Raking corralled berries into the elevator at My Achin' Back Bog, East Freetown. Susan Ashley (center left), Nancy Faulkner, Lynn Almeida and Jonathan Ashley. (John Robson photograph)

Milestone Bog, Nantucket. This 280-acre bog is the largest contiguous (shore to shore) cranberry bog in the world. Built in 1900 by George MaGlathlin, it is now owned by the Nantucket Conservation Foundation and leased and managed by Russell Lawton. By right, the grounds around the bogs are open to the public, and tourists generally picnic or stroll along the shores watching the workers harvest. With experienced growers managing the bog, the Conservation Foundation is earning far more by receiving a percentage of the crop than they did trying to run the bog themselves. (Joseph D. Thomas photograph)

Water-harvesting at My Achin' Back Bog, East Freetown. Cranberry grower Steve Ashley and his dog, Boy, walk ahead of Steve Bottomley (on the water reel) looking for muskrat holes and ditches that might disable the machinery. Boy does most of the work and Steve is learning from him constantly. (John Robson photograph)

Top left (Ted Young) and right (Carver bog): The dry-harvesting machines are pushed along the bog in a circular path, beginning on the bog's perimeter and working toward the center. The vines are trained to grow in the direction they are harvested and must be harvested the same way every year. (Joseph D. Thomas photographs)

Corralling berries at Poquoy Brook Bog, Ocean Spray, Lakeville. (Joseph D. Thomas photograph)

Young Scottish workers on a Harju bog in Carver. Many cranberry growers today are looking toward the British Isles for their workers. "We have six guys from Scotland," says Paul Harju. "They're good workers. There's not enough farm people around here. But we have to put up with too much government baloney to get them; too much paper work. For instance, we made a house for them, but because the beds weren't high enough off the floor, we had to put in new ones. They [the government] even check the dishes for cracks. Still, it's worth the hassle." (Beverly Conley photograph)

Cleaning wet-harvested berries at Milestone Bog, Nantucket. On Nantucket nearly the entire crew are Scotsmen. "They're good farm hands, they understand machinery—check the oil and the engine every day," says foreman Tom Larabee, who has worked the Nantucket bogs for 30 years. "They're not afraid to get a shovel and use it." (Joseph D. Thomas photograph)

Carrying a hoop frame, New Piece Bog, Carver. (Beverly Conley photograph)

Tasting the crop, Chipaway Bog, East Freetown. (John Robson photograph)

Corralling berries at Kallio bogs, Carver. (John Robson photograph)

Detrashing the harvest at Merry Bog, North Carver. Betty Cole directs a hose onto the berries as the conveyor brings them to the top of the elevator and dumps them into the truck. The leaves and twigs brought up with the berries are rinsed away and pass down a chute (foreground) that deposits them onto a bog buggy rigged as a "trash truck." (Joseph D. Thomas photograph)

Coralled berries at Wankinco. (Joseph D. Thomas photograph)

Berries beaten and knocked loose by the water reel mounted on David Mann's air boat, Plymouth. (Joseph D. Thomas photograph)

Larry Cole, helping corral berries at Merry Bog in North Carver. (John Robson photograph)

Harvesting at David Olson's Black Brook Bog in Carver. (Joseph D. Thomas photograph)

Frog at Wankinco. (John Robson photograph)

Near day's end at White Springs Bog, Wankinco. The job is not done until the berries are put on the truck. (John Robson photograph)

Harvesting at Windswept Bog, Nantucket. Water-reel harvesting machines are called "water beaters" by growers because the reel, usually rotating away from the machine, whips through the vines and beats the cranberries loose. (Joseph D. Thomas photograph)

Top left: David Mann's air boat, Garland Bog, Plymouth. Mann had this boat built in Florida patterned after the Everglades air boats (one of which he also owns). This boat was built to Mann's specifications—larger to carry more weight while drawing less water than its Everglades prototype. Other advantages of water-beating by boat are that the bog ditches are not an obstacle and the vines are spared the trampling and tearing caused by the harvester's tires. However, because the airboat takes longer to get the job done, its economic impact has not yet been determined. "We'll know for sure in a couple of years when we compare yields," says Mann. (Joseph D. Thomas photograph)

Water-harvesting at My Achin' Back Bog, East Freetown. Steve "Runner" Bottomley gets some relief from the sun while working on Steve Ashley's bog. (John Robson photograph)

Looking for muskrat holes and ditches, Merry Bog, North Carver. Larry Cole, with potato digger and weed hook, searches for the pitfalls of the water-reel harvester while pulling out nasty weeds along the way. His son Jim is operating the harvester. (Joseph D. Thomas photograph)

Harvesting at Poquoy Bog, Ocean Spray, Lakeville. Matthew Beaton leads Richard Gast and Eddie Morales along the beaten path of the water reel. These machines, known as Bailey Beaters, were built for Beaton Cranberry Growers' Service by John Bailey of West Wareham. They differ from some water reels in that the reel rotates away from the machine as it moves forward. Waterreels or "water beaters," can generally work in water 2 inches to 2 feet deep. These machines are 8 feet wide. (Joseph D. Thomas photograph)

Jonathan Ashley, corralling berries at My Achin' Back Bog, East Freetown. (John Robson photograph)

Above and left: Raking berries into the elevator, White Springs Bog, Wankinco, South Carver. Berries moving up the elevator and into the trailer truck are cleaned of some of their "trash" by a spray coming through nozzles on the elevator. (Joseph D. Thomas photographs)

Pumping berries off the bog, Kallio Bogs, Carver. Some growers are now using pumps instead of elevators to take the berries off the bog. The berries are pumped into a flume or tank and, along with the trash, float to the surface. They pass onto a grate where they are sprayed with water to screen out the trash. This method, though time consuming, cleans the berries more thoroughly than the elevator and deposits the trash more effectively. (Joseph D. Thomas photograph)

Top right: Taking berries off the bog using an elevator, White Springs Bog, Wankinco. When the berries are deposited from the elevator into the truck, they are spread out evenly in the trailer. The man spreading them blows his whistle for the truck driver to move a few feet forward in order for the berries to be evenly scattered. (Joseph D. Thomas photograph)

Cleaning berries on Nantucket. Berries are brought to the Milestone Bog's receiving station and are dumped into a giant hopper and cleaned along two conveyors before being deposited in another truck that will take them to the ferry for the mainland. (Joseph D. Thomas photograph)

Receiving dry-harvested berries at Betty's Neck screenhouse, Lakeville. When the 10-barrel bins come off the bog they are unloaded at one of Decas' screening operations. At Betty's Neck workers dump the berries onto a platform (top left), run them through an air duct (center), which removes some of the chaff, and separate them into the 1/3-barrel boxes so they can be handled more easily in the screening room. (Joseph D. Thomas photograph)

Top left: Dumping berries into a separating machine at Betty's Neck screenhouse, Lakeville. Decas Cranberry Company uses the same, vintage-1920 cranberry separators that Massachusetts growers have used to winnow and grade berries for six decades. (John Robson photograph)

Screening berries at Betty's Neck screenhouse, Lakeville. During the harvesting season, at least two shifts of workers attend the screening operations. This crew will be on until midnight. (Joseph D. Thomas photograph)

Repairing a Furford picker at the Decas maintenance shop at Mary's Pond, Rochester. The Decas Company prefers Furford over Darlington pickers because they use easy-to-handle burlap bags to hold the harvested berries, while most Darlingtons are equipped with the more cumbersome 1/3-barrel boxes. (John Robson photograph)

Interior of the Hayden Manufacturing Company, owned and operated by Robert St. Jacques, West Wareham. One of the industry's top suppliers and manufacturers of machinery and parts, old or new, the Hayden Company was founded in 1892 by Lothrop Hayden and bought by Emile St. Jacques in 1926. Operating from the old Tremont screenhouse of the New England Cranberry Sales Company, Hayden is the sole manufacturer of the Darlington picker. It also makes water reels, pumps, weed wipers, pruners, dusters, etc., as well as builds and installs irrigation systems. Seen in this photograph, hanging from the ceiling, are the Hayden insect nets. (John Robson photograph)

The remains of a boghouse along Route 58 in South Carver. Typical of most boghouses, the original structure was 12 x 10 feet and 6 feet high. Inside there was room enough for a couple of bunks and a wood stove; outside (to the right) was the pump for drinking water. This house was also used as a sauna by its Finnish owners. A stove and hearth made of rocks was built inside; a wooden keg used as a holding tank supplied water that circulated through pipes over the hot rocks, creating steam. Many Finns attribute their good health and endurance to their saunas. (Joseph D. Thomas photograph)

Housing at the Federal Furnace Cranberry Company, South Carver. This is probably the last vestige of a cranberry bog village with intact workers' housing. Strict housing standards for migrant workers in Massachusetts have made shanties a remnant of the past, and most have been torn down. (Joseph D. Thomas photograph)

Six growers, southeastern Massachusetts. Pictured here are: top, Anne Kallio and Robert Johnson; bottom, Joseph Barboza, Kirby Gilmore, Clark Griffith; opposite page, Robert Hammond. Anne Kallio and Robert Johnson are second-generation Finns who continue to employ the same high cultural practices set by their parents. Clark Griffith's family has harvested cranberries in South Carver for over 100 years, and he is well known for his community involvement as well as for his work ethics. Joey Barboza of West Wareham represents the many Cape Verdeans who worked their way into the industry from the ground up. He came into growing via the construction trade and bog building. Kirby Gilmore is the new generation of cranberry growers, exploring new ways and applying fresh ideas to cultivation and maintenance, while retaining many traditional practices passed down to him from other growers. (Photographs by John Robson, Joseph D. Thomas and Beverly Conley)

216

Robert Hammond is a patriarch among cranberry men. His knowledge of growing is highly regarded. One grower said of Bob Hammond, "He rarely needs the frost warning system, because he can feel it in his bones. He has his own equations—his instincts." (Joseph D. Thomas photograph)

Selected Bibliography

Books

Benson, Adolph B. *Peter Kalm's Travels in North America*. Dover Publications, Inc., New York (1770).

Burrows, Fredrika A. *Cannonballs and Cranberries*. William S. Sullwold, Taunton, 1974.

Buszek, Beatrice Ross. *The Cranberry Connection: Cranberry Cookery*, 2nd ed. The Stephen Greene Press, Brattleboro, Ver., 1978.

Child, Lydia Marie. *The American Frugal Housewife*. Carter, Hendee and Co., Boston, 1832.

Deyo, Simeon L. *History of Barnstable County, Massachusetts 1620–1890*. H.W. Blake & Co., New York, 1890.

Dodoens, D. Rembert. *A Nievve Herball, or Historie of Plantes*. trans. Henry Lyte, Gerard Dewes, London, 1619 (1578).

Eastwood, Benjamin. *The Cranberry and its Culture*. Orange-Judd & Company, 1856.

Eliot, Jared. *Essays upon Field Husbandry in New England*. Harry J. Carman and Rexford G. Tugwell, eds., AMS Press, Inc., New York, 1967 (1766).

Eliot, John. "The Day-Breaking, if not the Sun-Rising of the Gospel with the Indians in New England," in *Massachusetts Historical Society Collections*, 3rd series, IV. 1834 (1647), pp. 1-23.

Fosdick, Lucian J. Bog *Building and Cranberry Culture*. Ed. Freeman & Sons, Providence, 1902.

Garside, E. *Cranberry Red*. Little, Brown and Company, Boston, 1938.

Gomes, Peter J. "Plymouth and Some Portuguese," in *They Knew They Were Pilgrims: Essays in Plymouth History*, L.D. Geller, ed. Poseidon Books, Inc., New York, 1971.

Griffith, Henry S. *History of the Town of Carver*. E. Anthony & Sons, Inc., New Bedford, Mass., 1913.

Halter, Marilyn. "Working the Cranberry Bogs: Cape Verdeans in Southeastern Massachusetts," *Spinner*, Vol. III. Spinner Publications, Inc., New Bedford, Mass., 1984.

Holmes, O.M. "Cape Cod Cranberry Methods," from *Proceedings of the 13th Annual August Meeting of the American Cranberry Growers' Association* (Taunton, Mass.). 1883.

Josselyn, John. *New-Englands Rarities Discovered*. Mass. Historical Society, Boston, 1972 (1672).

"Memoir on the Consumption of Cranberry Sauce, by the Americans." reprinted in *Cranberries* magazine, September 1936 (1808).

Penti, Marsha. "The Life History of a Southeastern Massachusetts Finnish Cranberry Growing Community," in *Finnish Diaspora II*, Michael Karni, ed. The Multicultural History Society of Ontario, Toronto, 1981.

Report of the Commission on Immigration on the Problem of Immigration in Massachusetts. House Document No. 2300. Wright & Potter Printing Co., Boston, 1914.

Reynard, Elizabeth. *The Narrow Land*. The Chatham Historical Society, Inc., Chatham, Mass, 1978 (1934).

Shephard, Thomas. "The Clear Sun-shine of the Gospel," *Massachusetts Historical Society Collections*. 3rd series, IV. Boston, 1834 (1648).

Simmons, Amelia. *American Cookery*. Dover Publications, Inc., New York, 1958. (1796).

Sullivan, Philip. *Cranberry King*. Unpublished thesis, Williams College, Amherst, Mass., 1971.

Urann, Marcus L. *A Word Fitly Spoken*. privately published, 1943.

Webb, James. *Cape Cod Cranberries*. O. Judd Co., 1886.

White, Joseph J. *Cranberry Culture*. Orange Judd & Co., 1870.

Williams, Roger. *A Key into the Language of America*. Wayne State University Press, Detroit, 1973 (1643).

Winslow, Edward. "Good News From New England," in *Chronicles of the Pilgrim Fathers*. Alexander Young, ed. Da Capo Press, New York, 1971 (1624).

Periodicals

"Berry Good? Bounce Speaks for Itself," *Science News*, November 17, 1984.

"Cape Cod Cranberry," *Harpers Weekly*, October 9, 1897.

"Cranberry Culture," *The Land We Love*, December 1867.

"The Cranberry Industry," *Scientific American*, March 22, 1902.

"Cranberry Scare: Here Are the Facts," *U.S. News and World Report*, November 23, 1959.

Hall, Clarence. *Cranberries*. Sixteen-part series on the history of cranberries, December 1948-April 1949.

Hall, L.C. "The Cranberry Industry," *Cape Cod Magazine*, September 1923.

Lovejoy, Owen R. "The Cost of Cranberry Sauce," *Survey* magazine, January 7, 1911.

MacCulloch, Campbell. "Who Picked Your Cranberries?" *Good Housekeeping*, November, 1913.

The Naturalist. Vol. 1, No. 10, October 31, 1841.

The New England Farmer and Horticultural Journal, Vol. 10, No. 44 May 16, 1832.

Selected *New York Times* articles (listed by date):

"Cranberry Culture: J. Webb's Success in New Jersey," October 4, 1885.

"Cranberry Canners Inc. Organized by Cape Cod and New Jersey Packers," March 9, 1930.

"Man Wounded, 64 Strikers arrested," So. Carver, September 14, 1933.

"On Improved Technology: Water Harvesting Techniques," November 2, 1975."

"The Cranberry Industry of Southeastern Mass.: Produces Half the Entire Crop Produced," November 23, 1988.

"Ocean Spray's Juicy Future," *Business Week*, November 23, 1981.

Pamphlets and Reports

Bursley, John. *Cranberry Culture in Southeastern Massachusetts*, 49th Annual Report of the Secretary, Massachusetts State Board of Agriculture,.Wright & Potter Printing Co., Boston, 1902.

Corbett, L.C. *Cranberry Culture*, Farmers' Bulletin No. 176. U.S. Department of Agriculture, Washington, 1903.

Franklin, Henry J. *Cranberry Growing in Massachusetts*, Extension Leaflet No. 72. Massachusetts State College (University of Massachusetts), Amherst, Mass., 1923, 1940, 1948.

"Grades and Brands of the New England Cranberry Sales Company," November 1940.

Quarles, John R. *Ocean Spray: 1930-1960*. Address delivered at the annual stockholders' meeting, August 17, 1960.

Shear, C.L. *Fungous Diseases of the Cranberry*, Farmers' Bulletin No. 221. U.S. Department of Agriculture, Washington, 1905.

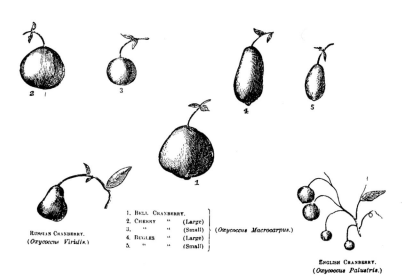

THE KNOWN CRANBERRIES OF COMMERCE.
Natural Size.

Index

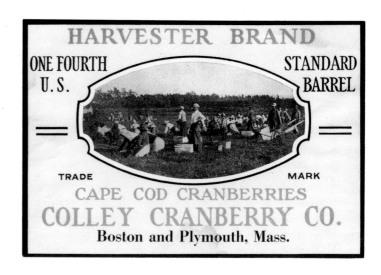

HARVESTER BRAND
ONE FOURTH U.S. — STANDARD BARREL
TRADE MARK
CAPE COD CRANBERRIES
COLLEY CRANBERRY CO.
Boston and Plymouth, Mass.

Spinner Publications records the history and culture of the cities and towns of southeastern Massachusetts. Spinner tells the story of the individual, the community and working the land. Understanding the area's history is essential in adjusting to current changes and making our way into the future. Spinner would like to contribute to this understanding by recording history in a way that is accessible to many people.

Studying family history brings history to a personal level and allows us to see how people have established homes, earned a living and participated in culture, all within a shaping context of geography, urbanization and industrialism.

Since 1980, Spinner has produced four volumes of the award-winning book series *Spinner: People and Culture in Southeastern Massachusetts*. We seek to promote the arts of the region and collaboration among artists to present local history in an accurate, dramatic, entertaining way. For information about our books, please write:

Spinner
Publications
P.O. Box C-801
New Bedford, MA 02741